THE QUANTUM LABYRINTH

Also by Paul Halpern

Einstein's Dice and Schrödinger's Cat: How Two Great Minds Battled Quantum Randomness to Create a Unified Theory of Physics

Edge of the Universe: A Voyage to the Cosmic Horizon and Beyond

Collider: The Search for the World's Smallest Particles

Brave New Universe: Illuminating the Darkest Secrets of the Cosmos

The Great Beyond: Higher Dimensions, Parallel Universes, and the Extraordinary Search for a Theory of Everything

THE QUANTUM LABYRINTH

How Richard Feynman and John Wheeler Revolutionized Time and Reality

Paul Halpern, PhD

BASIC BOOKS
New York

Basic Books
Hachette Book Group
1290 Avenue of the Americas, New York, NY 10104
www.basicbooks.com

Printed in the United States of America

First Edition: October 2017

Published by Basic Books, an imprint of Perseus Books, LLC, a subsidiary of Hachette Book Group, Inc.

Print book interior design by Amy Quinn.

Library of Congress Cataloging-in-Publication Data

Names: Halpern, Paul, 1961–author.
Title: The quantum labyrinth : how Richard Feynman and John Wheeler revolutionized time and reality / by Paul Halpern, PhD.
Description: New York : Basic Books, an imprint of Perseus Books, LLC, a subsidiary of Hachette Book Group, Inc., [2017] | Includes bibliographical references and index.
Identifiers: LCCN 2017013259 (print) | LCCN 2017017848 (ebook) | ISBN 9780465097593 (ebook) | ISBN 9780465097586 (hardcover) | ISBN 0465097588 (hardcover) |
ISBN 0465097596 (ebook)
Subjects: LCSH: Quantum theory. | Time. | Space and time. | Reality. | Feynman, Richard P. (Richard Phillips), 1918–1988. | Wheeler, John Archibald, 1911–2008.
Classification: LCC QC174.12 (ebook) | LCC QC174.12 .H347 2017 (print) | DDC 530.12--dc23
LC record available at https://lccn.loc.gov/2017013259

LSC-C

10 9 8 7 6 5 4 3 2 1

Dedicated to my brothers, Rich, Alan, and Ken

How come time? It is not enough to joke that "Time is nature's way to keep everything from happening all at once."

—John A. Wheeler, *Time Today*

I thought of a labyrinth of labyrinths, of one sinuous spreading labyrinth that would encompass the past and the future and in some way involve the stars.

—Jorge Luis Borges, "The Garden of Forking Paths"

CONTENTS

INTRODUCTION: A REVOLUTION IN TIME

> Tyger Tyger burning bright,
> In the forests of the night:
> What immortal hand or eye,
> Dare frame thy fearful symmetry?
>
> —William Blake, "The Tyger"

It is nighttime in Princeton, and we are going on a ghost hunt. The town is eerily quiet; all the shops are closed. A cold, full moon illuminates the leafy university campus.

More than seventy-five years ago, roughly corresponding to the start of World War II, a quiet revolution began here in our understanding of the nature of time. Discussions between two brilliant physicists, Richard Phillips "Dick" Feynman and John Archibald "Johnny" Wheeler, set off a chain of events that fundamentally recast the notion of time and history in quantum physics. Ultimately, their ideas transformed the concept of time from a single stream flowing inalterably in one direction into a labyrinth of alternatives extending backward as well as forward. By probing Princeton's past, we wish to unravel how this radical change was born and understand its impact on the contemporary search for a complete explanation of physical reality.

We start our trek into scientific history at Nassau Hall, the university's traditional hub. Bronze tigers, one on each side, guard its front entranceway in a wonderful spatial symmetry. Walking northward, we pass through FitzRandolph Gate, the campus's ornate portal with twin stone eagles perched on two monumental columns. We reach Nassau Street, Princeton's main thoroughfare—the demarcation between town and gown.

Gazing across the street, in contrast to the elegant architectural balance of the campus buildings, we notice a marked asymmetry. Eastward on the right is Lower Pyne, a marvel of Tudor gingerbread design, fashioned after sixteenth-century houses in Chester, England. It is truly stunning. Westward, on the left, is an unadorned bank. Austere, boxy, and cold, it seems an unworthy companion to the friendly, delicate edifice on the right.

We cross the street and are swept up in an unexpected haze. The clear night has suddenly turned foggy. Like a phantom in the mist, we see Upper Pyne, Lower Pyne's long-lost companion. Built in similar style at the same time, its most prominent feature is a sundial clock with the Latin motto *Vulnerant omnes: ultima necat* (The hours all wound, but the last one kills!). The building was demolished in the early 1960s to make way for the bank. But to our tired eyes at least, it seems to be standing just fine. Symmetry has been restored.

Palmer Square, even farther west, appears verdant and new. Its shops were built during a gentrification phase of the late 1930s. Strangely, they look like they've just opened. A newsstand displays a headline about Adolf Hitler's invasion of Poland—which we remember happened in September 1939. A movie poster advertises *The Wizard of Oz*. I've a feeling we're not in the twenty-first century anymore.

THE GRADUATE

After a bit more walking we find ourselves in Princeton's Graduate College, a castle-like enclave just off the main part of campus. The complex is a cloister within a cloister, offering an isolated environment for busy graduate students. Here, students reside in simple but comfortable dorms, take their meals in a central dining hall, and attend posh social functions such as dances and teas.

Most of the residents are asleep. But lights are on in a small, ornate library room, where a lanky, brown-haired twenty-one-year-old, slouched in a chair, mouth curled into the hint of a smile, gazes intently at a book on classical mechanics propped on his lap. He is a first-year graduate student preparing for an undergraduate course for which he will serve as a teaching assistant and grader. While the material is familiar to him, he has decided to take a quick look at what might lie ahead in the course. He is bracing for the imminent challenge of wading through piles of homework, checking students' calculations and correcting their errors in a way that motivates them to hone their problem-solving skills.

A pyramid-shaped table lamp illuminates the passage the young graduate student is reading. It is about the head-on collision of two carts on a frictionless track. He runs through the problem in his head. Given the carts' masses and initial velocities, the laws of physics dictate exactly what will happen next. According to Isaac Newton's third law of motion, for every action there is a reaction of equal magnitude and opposite direction. That means each cart experiences the same amount of force due to the other, but aimed in opposite ways. Following Newton's second law, force is change of momentum: the product of mass times velocity. Because each cart feels the same force, it changes its momentum by the same amount: one giveth and the other taketh. That universal balance is called the "law of conservation of momentum."

With perfect symmetry, the carts move away from each other with the same momentum boost, but in opposite directions. What about their speeds? Well, given that momentum constitutes mass times velocity, the lighter one will zoom off faster than the heavier one. That's the beauty of classical Newtonian physics. ("Classical" in this context refers to the familiar scale of everyday life, as opposed to the subatomic "quantum" scale.) We can make a precise prediction through a simple conservation law.

Elsewhere the book includes a section on simple harmonic motion: the behavior of springs, rubber bands, pendula, and anything that snaps back toward equilibrium (the balanced position) when stretched, compressed, or swung. Springs are often used to represent any such elastic object. Just as in the case of collisions, classical principles guarantee that spring motion is completely predictable. Ignoring friction, a spring stretched and let go will find its way back to its unstretched position. By the time it reaches that unstretched point, called "equilibrium," it will be moving at its maximum speed. That is because its energy recycles from one form to another. The energy associated with its initial position, called "potential energy," transforms into energy connected with its motion, called "kinetic energy." But the drama doesn't end there. The spring keeps moving until it becomes compressed. At the point of maximum compression, it briefly stops and turns around. Its kinetic energy has completely changed back into potential energy—this time associated with squeezing, not stretching. The spring returns to equilibrium and keeps going until it stretches once again. The recycling of energy from potential to kinetic to potential to kinetic, and so forth, is called "conservation of energy."

A simple pendulum does the same thing. Back and forth, back and forth it swings—converting potential energy into kinetic and back into potential again. If only there were no friction, it would keep on swinging forever. A clock or watch, for that matter, could keep ticking forever in that ideal situation. It is a perfect, perpetual rhythm, fixed by the metronome of conservation laws.

The young scholar starts banging out a simple beat on the table next to him. Thump, thump. Thump, thump. Thumpety, thump, thump, thump. All is rhythm.

Cyclical time, the notion that time is repetitive and that certain patterns of events happen again and again, follows directly from nature's recycling of mechanical energy. Closed systems that perfectly conserve such energy tend to repeat themselves over and over. For complex arrangements, such cycles might be astronomically long. Nonetheless, finite systems that recycle their energy eventually recur—as when playing tic-tac-toe indefinitely one eventually must repeat the same moves. Nature loves rhythmic patterns.

Yet other types of energy cannot be completely reused, such as the heat generated by engines due to friction or air resistance. That buildup of waste energy engenders a natural arrow of irreversibility pointing to the future. Consequently, while some ideal systems follow a kind of cyclic time, many realistic physical processes obey a linear time scheme. The question of cycles versus linear arrows has been at the core of discussions of time for millennia.

The graduate student yawns. His desk drumming tapers off. The book falls to the floor. Suddenly commanded to slumber by his own internal clock, he stands up, shambles toward his dorm room, and tumbles into bed. He'll need his sleep; in the morning he's scheduled to head to a building called Fine Hall to meet with his advisor. Daybreak will bring his marching orders for his duties as a teaching assistant.

QUANTUM PROFILES

Fine Hall (now called Jones Hall) is about a mile eastward across Princeton's campus from the Graduate College—an easy walk for an energetic young student. Constructed specifically for the math department, it features thick, leaden windows artfully adorned with mathematical symbols. In the fall semester of 1939, it housed the offices of several theoretical physicists, including Eugene Wigner and John Wheeler. Until the spring of that year, it served additionally as the home of the Institute for Advanced Study (IAS), an independent think tank that counted physicist Albert Einstein, Hungarian mathematician John von Neumann, Austrian mathematician Kurt Gödel, and many other notables among its original members.

For Einstein, the institute's most famous researcher, the IAS served as kind of a cloister where he was free to pursue his dream of a unified theory of gravitation and electromagnetism, while offering an iconoclastic critique of probabilistic quantum mechanics: the physics that applies to atoms and subatomic particles. His persistent arguments against quantum "dice rolling" and belief in pure determinism isolated him from the mainstream physics community. Determinism in this context means that if one knew perfectly all the initial conditions of a physical system, such as a pendulum or spring, one could predict exactly what would happen indefinitely into the future. Einstein aspired to "complete" quantum mechanics by eliminating chance aspects that arise in taking measurements.

Von Neumann, in contrast, developed a more nuanced view of quantum mechanics in which determinism and chance each plays a role during different stages. In his classic 1932 textbook, *Mathematical Foundations of Quantum Mechanics*, he presented a two-step analysis of quantum processes. Before an experimenter takes a measurement of a quantum system, such as an electron in an atom, its dynamics flow smoothly and predictably. However once he switches on his measuring device—a strong magnet, let's say—and takes a reading, chance kicks in, and the outcome might be one of many—selected as randomly as coin flips. Why should the observer play such a pivotal role? What does it mean to observe? Might anyone or anything be an observer? Could the observer be part of the system itself? These questions fall into the purview of what is called the "quantum measurement problem."

The issue of quantum measurement is a tricky one. Unlike in classical mechanics, in quantum mechanics one cannot access all of the information about a particle directly—that is, its position, velocity, and so forth. Rather one must consider an entity called the "wave function," which contains all the data about a particle's quantum state. Rather than containing exact values, it offers a probabilistic spread showing the chances that the particle will respond a certain way if measured. (Technically the *square* of the wave function yields the probability distribution.) Peaks represent higher odds; troughs offer lower chances. It is like a bell curve showing that if you flip four coins, the most likely combinations have two heads and two tails, in any order, and the least likely have either all tails or all heads.

As von Neumann pointed out, the wave function undergoes two separate types of quantum processes: continuous evolution via Schrödinger's wave equation and discrete "collapse" whenever an observer takes a measurement. For example, suppose an observer conducts an experiment designed to record an electron's exact location. Before the observation, the electron's wave function would continuously obey Erwin Schrödinger's equation telling it exactly how to behave. Nothing would be left to chance. Immediately after the observation, however, the wave function would randomly collapse from a smooth probability spread into a sharp spike representing a particular position value.

While the first kind of process is wholly deterministic and reversible, the second is random and irreversible. These embody different conceptions of time: the first mechanism matches the cyclic time of a classical pendulum or spring, and the second embodies the linear, irreversible time of an engine wearing down and ultimately grinding to a halt.

By the late 1930s, von Neumann's dual picture of continual, reversible evolution, followed by instant, irreversible collapse, had become the orthodox view of quantum measurement—what has come to be known as the "Copenhagen interpretation." Awkwardly, however, it embraced an odd combination of cyclical and linear time that didn't really mesh—like an otherwise perfect watch that stopped abruptly and irreparably whenever you happened to look at it. Observation would break the mechanism—unacceptable in a Rolex but deemed credible in quantum mechanics. Because experimental data beautifully matched theory, most scientists simply accepted the bizarre idea that observation changed the dynamics of a quantum system from predictable continuity to random jumps. Only a few notable critics, such as Einstein, Schrödinger, and Louis de Broglie (who had developed the original idea of matter waves that motivated Schrödinger's wave equation), called for a rethinking of the scheme.

A WONDERFUL FREAK OF FATE

In spring 1939, the IAS relocated to a verdant new campus. Einstein, von Neumann, and other members moved to comfortable new offices in colonial-style Fuld Hall. Vacated by the institute members, Fine Hall lost many prominent thinkers. Yet within its ivy-covered walls a revolution would begin that offered a third way of viewing time: beyond cyclical and linear. The new approach, which would be called "sum over histories," presented time as a labyrinth of alternatives.

Did determinism or chance lead young Richard Feynman to Princeton, where he would live in the Graduate College and work with John Wheeler in his Fine Hall office and the adjacent Palmer Laboratory physics building? It was a monumental match of highly original thinkers who had the gumption to rebuild quantum physics from the ground up, based on novel principles.

When admitted to Princeton, Feynman had originally been assigned as Wigner's teaching assistant. Wigner was a Hungarian physicist who came to share a passion for quantum measurement theory and held views on the subject similar to von Neumann's. At the last minute, Feynman was switched to assisting Wheeler instead.

In retrospect, each considered the substitution one of the most auspicious moments in his career: "Through some wonderful freak of fate I ended up with him assigned to me," Wheeler later recalled.

"I was very lucky when I got to Princeton . . . and was Wheeler's research assistant," Feynman said. "You might say that my success was a result of things I learned from him."

Feynman and Wheeler's collaboration, as it turned out, would lead to a rethinking of the fundamentals of quantum physics through the concept of "sum over histories," introduced by Feynman and named by Wheeler. That revolutionary approach sees actuality as a composition of all possibilities, like a song with multiple tracks blended together. How does an electron cross the road? As Feynman and Wheeler showed, the correct quantum answer is that it takes every physically possible path—with reality a combination of them all.

The two physicists would prove the perfect team: Feynman cautious and thorough in his brilliant calculations, Wheeler bold and imaginative in his far-reaching notions. Honing and reworking bizarre hypotheses into workable solutions would become their joint specialty. A lifetime journey of intrepid explorations would launch in Wheeler's Princeton office.

THE OUTSIDER

Richard Feynman was resolutely an outsider at Princeton—as if he had arrived from another planet. Born on May 11, 1918, to a secular Jewish family in New York City and raised in the borough of Queens, he had an unpolished, working-class accent (similar to Brooklynese) and unvarnished manners that marked him as strikingly different from the white, male, Protestant, wealthy prep school students (at least the undergraduates) who filled the university's halls at the time. Some in such a position would want to fade into the background and keep quiet. Certainly not Feynman, who realized from an early age that life is too short and time too precious to base one's actions on what other people think. He knew that he stood out but found this a source of humor and strength rather than embarrassment.

"Princeton has a certain elegance," Feynman later recalled. "And I was not an elegant person. In any formal social situation, I was really quite a clunk. . . . I was kind of a rough, kind of a simple character, as far as society goes. But I wasn't worried about it. I was just sort of half-proud of it."

On his very first day on campus, he had noted the posh speech and pretentiousness that marked the genteel class around him. He flinched at the academic gowns students needed to wear to social gatherings. Barely an hour after his father, Melville Feynman, dropped him off and left, a pompous "master of residence," with an affected upper-crust English accent, greeted him in his dorm room and invited him to afternoon tea with the dean of the Graduate College, Luther Eisenhardt. Feynman would have felt much more at ease at Nathan's hot dog stand on Coney Island, but he had never been to a formal tea before and was very curious.

Much like Margaret Dumont in the Marx Brothers' movies, the dean's wife wrapped herself in decorum. Dutifully, she extended greetings to each of the incoming students and offered them tea with either milk or lemon. She approached Feynman, who was pondering where to sit, with that question. Absentmindedly he replied, "I'll have both, thank you." Her bemused response, "Surely you're joking, Mr. Feynman!" accompanied by a titter of nervous laughter, would become the punch line of a much-repeated anecdote and later the title of his most popular book. Indeed Feynman's accent and antics would later inspire writer C. P. Snow to quip that it was "as though Groucho Marx was suddenly standing in for a great scientist."

Feynman's mission wasn't to conform to the styles and whims of the Ivy elite—or indeed to the expectations of anyone. Rather, he was passionately curious about the world and saw Princeton as a place that might equip him with the tools for unlocking its secrets. In particular, he had heard about its phenomenal cyclotron, a machine in the basement of Palmer Lab that used powerful magnets to whirl particles around in tight circles, voltage boosts to speed them up with each cycle, and targets for them to smash into once they were sufficiently energized. The machine had yielded bountiful results so far, and Feynman had been excited to see it.

CURIOSITY'S CHILD

Feynman's father, who was in the uniform business by profession but had a strong amateur interest in science, had endowed him with an insatiable fascination with how things work at their very core. When he was a child—nicknamed "Ritty" during those early years—his father treated him to puzzle after puzzle, such as colorful tiles to assemble in certain patterns. They would often explore the wonders of nature, such as barnacles on the beach, and enjoyed poring through encyclopedia articles on various subjects.

Thanks to his dad's encouragement, Richard picked up integral calculus when he was only thirteen. By then, the family had moved to a comfortable house in the Queens neighborhood of Far Rockaway, close to a popular beach. Richard's mother, Lucille, who had once aspired to be a kindergarten teacher, supported her husband's emphasis on science and math but also encouraged her son's literary side by often sharing funny stories.

By secondary school, Far Rockaway High School, Richard was so advanced that one of his teachers, sensing his boredom, gave him a calculus book to read on his own during class. He learned about Pierre de Fermat's principle of least time, a natural way of explaining why light travels in a straight line, around then. He also became familiar with the concept of time as the fourth dimension and other advanced topics in physics. A first-place finish in a New York University interscholastic math contest in 1935 won him a gold medal and mention in the *New York Times*.

Feynman often thought about how to explain things to his dad, who regularly peppered him with deep questions about fundamental physics, even after he went to college. "What makes it go?" his father would often ask him about natural occurrences. For example, one summer after Feynman had begun his undergraduate education at the Massachusetts Institute of Technology (MIT), his father asked him to explain an atomic process by which an electron drops to a lower energy level by emitting a photon (light particle). "Is the photon in the atom ahead of time, so that it can come out?" Melville asked him. "Or is there no photon in the atom to start with?"

Feynman tried valiantly to explain to his dad that like a boy uttering word after word in ordinary speech and not using them up, photons are unlimited. Just as there's no maximum allotment of nouns that a person might say, there's no bag of photons inside an atom that might become exhausted over time. Much to Richard's disappointment, his father remained perplexed about what actually happened when an electron emitted a photon. Ironically, Feynman's later Nobel Prize–winning work would involve such interactions.

Influenced by his father, Feynman would retain a childlike awe of the world throughout his life. As his friend Ralph Leighton later described, "He can always look at something the way a child does. He sees things with curiosity and wonder, finding something new and making a little puzzle out of it to think about."

Even by the standards of one of the top universities in the world for science and technology, Feynman was a brilliant student. He was extraordinary at mental calculation and a powerhouse at integration and other techniques of calculus. In spring 1939, during his senior year at MIT, he was invited to participate in the prestigious Putnam Mathematics Competition as part of the university's five-person team. At first he balked, protesting that he wasn't a math major. Once he realized, however, that there weren't enough suitable seniors to fill the team, he consented. Much to his surprise, he scored the highest by far in the country that year. His victory made national news and garnered him automatic admission to Harvard and a full scholarship in its graduate program.

Feynman was inclined, at first, to continue at MIT for his graduate studies. However, John Slater, the physics department chair and an accomplished quantum theorist, urged him to look elsewhere. After Feynman insisted that MIT was the best place to do science and that he should stay there, Slater rebutted that other institutions were comparable and offered novel opportunities. In the end, he put his foot down and barred Feynman from continuing.

While Harvard would have been another excellent choice, Feynman selected Princeton. Along with hearing about its cyclotron, he was familiar with the quantum physics research of Wigner and hoped to work with him. It was a bit of a shock to him when, upon arriving, he learned that Princeton's administration had suddenly switched him to working for Wheeler instead. The decision proved pivotal to both of their careers.

BOYHOOD ANTICS

Born in Jacksonville, Florida, on July 9, 1911, Wheeler was less than seven years older than Feynman. Like him, Wheeler had been strongly influenced by educated, involved parents. Wheeler's dad, Joseph Wheeler, was a respected librarian who headed several different libraries around the country, including Baltimore's well-known Enoch Pratt Free Library, and advocated for and supervised the construction of many branches. As the family moved from place to place—including California, Ohio, Vermont, and Maryland—as his father switched jobs, one mainstay was that their household was always full of books. John's mother, Mabel (nicknamed "Archie"), also a librarian, loved to read and bestowed upon him a lifelong love of the printed word. As a child, he often barraged her with questions about the universe such as, "If I keep going out into space, will I ever come to an end?"

Though each would become a theorist, Feynman and Wheeler shared a childhood passion for hands-on experimental science. Each was ardently curious about how things work. Both loved chemistry sets, radios, motors, and electrical kits. Prompted by a remark from his father about the importance of electrochemistry, Feynman once wired up a pile of dry chemicals to see what would happen. Finding countless ways to tinker with household items, he set up a makeshift intercom system in the house and motorized his little sister Joan's crib so that it would rock automatically.

Wheeler, as a child, was similarly adept at building things from scratch. He constructed crystal radio sets, a telegraph between his house and a friend's, an adding machine, and a combination lock. He experimented with gunpowder and almost lost a finger igniting dynamite caps near a pig barn.

If Feynman and Wheeler had grown up together, they likely would have had many hours of mischievous fun finding creative ways to make things light up and chemicals explode. When they met as young adults, an undercurrent of boyish exuberance charged their interactions. They loved to tinker together with everything from mechanical equipment to the fabric of space and time itself.

A REMARKABLE SYMMETRY

The relationship between a PhD advisor and his student is often imbalanced. After all, the former has enormous power over the latter's career. A poor or malicious supervisor might dole out bad advice to a student, drag out his thesis work, block him from receiving his degree, and effectively prevent him from pursuing an academic career.

The case of Feynman and Wheeler offers the rare exception in which the student-mentor relationship blossomed into a genuine, egalitarian friendship. The kinship between the two physicists only strengthened over the years, as each nurtured the other's growth. Each was a bold, open-minded thinker prepared to entertain even the wildest of suggestions. Bizarre notions flowed from both of their creative minds—from particles traveling backward in time to parallel strands of reality; from a universe crafted from pure geometry to one based on digital information. Arguably, much of the visionary work in theoretical physics in the late-twentieth and twenty-first centuries derives from their bold discourse, including the basis of the Standard Model of particle physics and all manner of astrophysical concepts, such as the properties of black holes and wormholes.

Each kept in his heart a youthful vision of the world as a wonderful place to explore: full of puzzles to piece together, secret codes to decipher, hidden passages to map out, and riddles to solve. Like Tom Sawyer and Huckleberry Finn, they were unsatisfied with mundane doings and pined for adventure.

Some symmetries are not immediately obvious. Judging from their outward personas, it would be hard at first to sense Feynman and Wheeler's common perspective. While Feynman hated stodginess, often spoke off-the-cuff in unadorned language, and exuberantly defied public expectations of a "serious scientist," Wheeler was quiet, measured, and polite in his speech and actions. Openly, Feynman was certainly the more unconventional of the two. Yet Wheeler was in his scientific pursuits even less mainstream. Beneath his veneer of conformity raged a nonconformist. In tandem, they were unafraid to toss out the old textbook explanations and begin anew. Perhaps two words could best describe their interactions: "crazy ideas."

Starting with Feynman's unexpected appointment as Wheeler's teaching assistant for a mechanics class, their explorations took off from there. Formality dissolved in a flash as the two quickly bonded as explorers of extraordinary possibilities in physics. Ultimately they reshaped the concept of time itself, allowing for alternative realities and backward journeys.

CHAPTER ONE

WHEELER'S WATCH

> This chap from MIT: Look at his aptitude test ratings in mathematics and physics. Fantastic! Nobody else who's applying here at Princeton comes anywhere near so close to the absolute peak. . . . He must be a diamond in the rough. We've never let in anyone with scores so low in history and English. But look at the practical experience he's had in chemistry and in working with friction.
>
> —John A. Wheeler, on the Graduate Committee's reaction to Feynman's application to Princeton

John Wheeler took his watch out of his pocket and put it on the table. He wanted to clock the meeting with his new teaching assistant, Richard Feynman. For a young assistant professor with teaching responsibilities and research interests, time is of the essence. Lecturing takes time. The deep concentration needed to tackle fundamental questions in physics takes time. Marking papers takes time. Meeting with students takes time.

A clock was ticking globally as well. The Nazis were on the march and had to be stopped. If they continued their conquest of Europe, it was only a matter of time before the United States would find itself obliged to join the war effort. Combatting terrible weapons possibly being developed by the Nazis might require scientific breakthroughs. As Wheeler had learned in January 1939 via his mentor Niels Bohr and Bohr's assistant, Leon Rosenfeld, researchers in

Germany had discovered that massive uranium nuclei could divide under certain circumstances, releasing a store of energy in a process called "fission."

The "chain reaction" that led to the startling news had been swift. Austrian physicist Lise Meitner, who had worked with German chemists Otto Hahn and Fritz Strassmann on the fission discovery, had told her nephew Otto Frisch. Based at the time at the Institute for Theoretical Physics in Copenhagen, Denmark, Frisch conveyed the results to Bohr, its director. Immediately realizing the monumental significance, Bohr told Rosenfeld and decided to make an announcement during an upcoming conference on theoretical physics, to be held at George Washington University in Washington, DC, on January 26. However, during a meeting of the Princeton physics department's Journal Club, held on January 16, the day Bohr and Rosenfeld arrived in the United States, Rosenfeld spilled the beans, letting Wheeler and others at Princeton know about the fission discovery. When Bohr somberly announced the findings at the Washington conference, the gravity of his words resonated within the broader physics community.

Many physicists who learned of this development—particularly those who had recently fled fascist regimes in Europe—were horrified by the idea that the Nazis might develop a bomb that tapped energy released by splitting atomic nuclei. Those especially concerned about the prospect of a German arsenal of nuclear weapons included Enrico Fermi, who had left Benito Mussolini's Italy for the United States, as well as Eugene Wigner, Leo Szilard, and Edward Teller, all émigrés from Hungary. Two months after Bohr's announcement, Fermi met with navy officials in Washington, DC. That summer Szilard, accompanied alternatively by Wigner and Teller, alerted Albert Einstein, who famously sent a warning letter to President Franklin Roosevelt. Given the Nazi threat and the possibility of American involvement, who knew when the US government might entreat nuclear theorists to drop their abstract endeavors and switch to military research?

Through work with Bohr, Wheeler had become one of the world's foremost experts in nuclear fission and would likely be tapped for his knowledge in the case of American entry into the war. The pair's joint research had begun five years earlier, in the fall of 1934, when Wheeler visited Bohr's institute. Fresh from PhD studies at Johns

Informal portrait of John Archibald Wheeler at Niels Bohr's Institute for Theoretical Physics in Copenhagen, mid-1930s (*Source:* AIP Emilio Segre Visual Archives, Wheeler Collection).

Hopkins, under the supervision of Austrian American physicist Karl Herzfeld, and a postdoc at New York University, mentored by Gregory Breit, Wheeler was eager to crack the mysteries of the atomic nucleus. He saw an apprenticeship with Bohr, the revered sage of quantum physics, who attracted scientists from around the world, as an ideal way to gain expertise in the topic. Wheeler worked in Copenhagen until June 1935, focusing on the interactions between nuclei and cosmic rays (energetic particles from space).

Bohr's style of investigation had a powerful effect on Wheeler. Though notoriously soft-spoken and known to mumble, Bohr had a knack for asking penetrating questions that revealed novel ways of thinking about a subject. As Wheeler recalled, "Bohr has this probing approach to everything, wanting to get down to the brass tacks and just test the uttermost limits to which a thing can be defended."

After returning from Europe, Wheeler enjoyed a three-year stint at the University of North Carolina, Chapel Hill, before being appointed assistant professor at Princeton in the fall of 1938. Even before Bohr announced the German fission program, times for the region were scary. That Halloween, according to a radio report, a Martian invasion hit nearby Grover's Mill. The fake broadcast, recounted in the authoritative voice of Orson Welles, sparked a panic. The public's terrified reaction reflected widespread fear of horrific new weapons. When, several months later, Bohr alerted physicists at the Washington conference about the nuclear fission discovery in Germany and the concomitant possibility of the Nazis building atomic bombs, visions of devastating terror attacks occupied everyone's worst nightmares.

Bohr stayed in Princeton from January until May 1939, working in an office on the same floor as Wheeler's in the building then known as Fine Hall (and now called Jones Hall) and residing at the Nassau Club. Trying to determine the precise mechanisms for fission, the two men employed Bohr's "liquid drop model," a flexible picture of the nucleus as something like a distended egg yolk that if stretched far enough could divide. Working together throughout the spring of 1939, they carefully determined the conditions under which fission might transpire when a uranium sample is bombarded with either fast (high-energy) or slow (low-energy) neutrons.

Wheeler drew pictures of the energy barriers the neutrons would need to cross for various isotopes (nuclear types) of uranium in order to seed themselves in their nuclei and split them apart. He modeled these barriers as something like hills skiers would need to ascend before reaching a summit and beginning a speedy descent. For the most common isotope, uranium-238, the initial hill was very steep and required fast neutrons—something like Olympic skiers—to make the jump. For a rare isotope, uranium-235, the barrier was lower and crossable by slow neutrons—something like novice skiers. Therefore Bohr and he concluded that uranium-235 was much more readily fissionable than uranium-238. Moreover, they discovered that a certain artificial isotope called plutonium-239, yet to be produced, could similarly be split using slow neutrons.

Given that the fission process creates more neutrons, these, if slowed down, could induce other nuclei to decay, leading under certain circumstances to a chain reaction and the controlled production of energy—or perhaps a massive explosion. Bohr and Wheeler published their results in a seminal paper, "The Mechanism of Nuclear Fission," which appeared on September 1, 1939—precisely the date Adolf Hitler invaded Poland and World War II began in Europe. Their findings would later prove indispensable for the Manhattan Project, the American wartime program to develop a nuclear bomb.

By that fall, Wheeler had set this collaborative work aside and was eager to make his own mark in theoretical physics. He also hoped to be a trusted mentor, as Bohr was to him. In the ideal professorial mix, deep reflection and quiet calculations, the private side, precisely balanced pedagogy, the public side. Maintaining the symmetry required perfect timing—hence the watch on the table.

Only twenty-eight years old, Wheeler scarcely could have known that he'd have almost seven decades left on this earth to ponder perplexing questions such as "How come existence?" (as he would often ask in his later years). The older Wheeler might have advised his young self to relax and enjoy his interactions with the students. But as the second hand made its rounds again and again only to push the minute hand farther and farther ahead, the young Wheeler took the task of balancing his responsibilities quite seriously.

FINE SILLINESS

Wheeler's office at the time, Room 214, was on the second floor of Fine Hall. Named after Henry Burchard Fine, the founder of Princeton's mathematics department, who had died tragically in 1928 when hit by a car while riding his bike, the building had been endowed by his friend Thomas D. Jones as an ornate temple of mathematical research. This mission became extended to theoretical physics. Each office included oak paneling, a blackboard, built-in filing cabinets, and windows overlooking a leafy enclave of Princeton's campus. The crisp aroma of autumn met with whiffs of chalk dust as professors attempted to characterize the natural world outside through their scrawlings within. It was a splendid location to do fundamental research.

Jones, mathematician Oswald Veblen, and others had designed Fine Hall to be as collaboration friendly as possible. To that end, a cozy tearoom, where faculty could congregate and discuss ideas, capped the rectangular array of offices on its second floor. Carved above the tearoom's fireplace was an inscription in German taken from one of Einstein's lectures: "Raffiniert ist der Herrgott, aber boshaft ist er nicht" (the Lord is clever, but He is not malicious). The saying reflected Einstein's belief that though the search for the proper equations of theoretical physics might have many twists, turns, and dead ends, nature would be cruel not to provide an ultimate solution.

The corner stairwells and interconnected corridors in the building were well used. Professors and students often retreated to the third floor, where a spacious library held thousands of volumes related to mathematics and physics. Sometimes they headed down to the first

Princeton's Graduate College (*Source:* Photo by Paul Halpern).

floor to attend seminars in a central lecture hall. Or, as did Bohr and Wheeler when they worked together, they paced around the cyclic hallways of the second floor, deep in discussion. As intended, the structure pulsed with the circulatory flow of researchers: up, down, and around.

To promote collaboration between mathematicians and physicists, a bridge connected Fine Hall to Palmer Laboratory, the main physics building, where classes were held and research conducted. Given the need for ample room for equipment, Palmer was far more spacious. To inspire experimenters, statues of two titans of American physics—Benjamin Franklin and Joseph Henry—framed the building's main entrance.

Upon meeting Wheeler, Feynman first noticed how youthful he appeared. He was certainly no statue, preserved for appearances like professors retained long past their prime; rather he was very young and vital. Feynman felt very much at ease. Then he observed Wheeler take out the watch and put it on the table to time their meeting. They discussed Feynman's responsibilities and made plans to meet again.

Not quite sure what to make of the timepiece, Feynman decided to play the same game. He bought a cheap watch and prepared for their next encounter. The second time they met, as soon as Wheeler reached into his pocket, Feynman did the same. Feynman's watch immediately followed Wheeler's onto the table, like a countering move in a chess game.

Feynman's sly imitation shattered the seriousness of their relationship. Wheeler started laughing hard. Feynman did too. The fits of hysterical laughter seemed to go on and on, as each did his best to crack the other up. The meeting had devolved into pure silliness.

Finally Wheeler decided it was time to get back to business. "Look, we have to get serious here and get going," he said.

"Yes, sir," replied Feynman with a smirk, and the two guffawed some more. Time and time again, meeting after meeting, discussions led to jokes, breathless fits of laughter, gasping pleas for seriousness, and then back to creative discourse. Feynman was used to switch-hitting; his mom, Lucille, often joked around, and his dad, Melville, was scientific and serious. With Wheeler, Feynman could express both sides of his personality. The stage was set for a long, productive—but often silly—friendship.

THE MECHANICS OF TEACHING MECHANICS

Wheeler took much pride in teaching and delivered a well-organized class in classical mechanics. He'd assign challenging homework problems, which the students would complete and turn in. Then it was Feynman's turn to test the students' mettle. He'd peruse the papers meticulously, looking for signs of logical flaws or calculation errors, make detailed comments in the margins, and return stacks of painstakingly marked homework to his mentor. The students had little chance of getting away with careless work or misconceptions.

Overjoyed that his assistant was doing such an exemplary job, Wheeler even let Feynman deliver at least one of the mechanics lectures, offering him valuable teaching experience. Honored by the invitation, Feynman stayed up all night preparing. He wrote to his mother that he was proud of the lecture, which went "nicely and smoothly," and expected to do a lot of that someday. Under Wheeler's wing and later on his own, Feynman would develop into a renowned explainer of physical ideas.

One of Wheeler's hallmarks as a lecturer, which had a profound influence on Feynman, was his use of clever diagrams. When framing an idea, he would almost always start off with a sketch, delineating the players involved and their interactions, as if planning a football strategy. As he later related, "I don't know how to think without pictures."

Both physicists saw teaching a subject as the best way to learn it. That might seem paradoxical: How can you explain something if you are not an expert? True, in the case of something relatively static, such as Latin or ancient Greek, you need to be fluent before delivering a solid lecture. However, physics is built from the ground up, based on fundamental principles that might be stated or interpreted in many ways. Even concepts typically addressed in the first weeks of an introductory physics course, such as force and inertia, are nuanced.

Inertia is the concept that objects at rest remain at rest, but those in motion keep going at the same speed in the same direction, unless acted on by an external force. It is why a bowling ball rolling on a flat, frictionless surface keeps going in a straight line before striking pins. Strangely, it is not a force but rather the lack of force that keeps the ball heading steadily toward its target. Intuitively we think that a force must be doing this, but reality tells us otherwise. Helping students understand that distinction is an intellectual challenge that sets the mind thinking about other aspects of the physical world. Explaining such notions might reveal new connections that elucidate the fundamental workings of nature.

For example, planning the mechanics course motivated Wheeler and Feynman to discuss Mach's principle, the idea that the distant stars somehow cause inertia. In contrast to Isaac Newton, whose physics frames inertia in terms of abstractions called "absolute space" (fixed yardsticks) and "absolute time" (abstract clocks ticking the same everywhere for all times), physicist Ernst Mach proposed that inertia must have a physical cause. He conjectured that the combined tugs of remote bodies induce a stationary object to remain at rest and a moving object to maintain its velocity.

EINSTEIN'S COSMIC VISION

As Wheeler knew well, Einstein's general theory of relativity—his masterful set of equations describing gravitation—attempted to realize Mach's principle and toss aside the unphysical Newtonian notion of an invisible, absolute framework for measuring inertia. Newton imagined spatial distances and temporal durations as invariant from point to point and moment to moment, like the fixed coordinate axes used by mathematicians. Nothing physical can affect those inert yardsticks. Sharply contrasting with Newton's steely, permanent measuring rods, in general relativity matter and energy warp the fabric of spacetime (space and time combined), like a heavy nest on a flimsy tree branch.

Along with banishing absolute space and time, Einstein's use of geometry to explain gravity eliminated another conundrum in Newtonian physics, called "action at a distance": forces, such as gravitation, acting instantly and remotely. For any two massive objects, Newton imagined a kind of abstract "thread" linking them together through their mutual gravitation. Nothing tangible in space would serve as an intermediary.

In Newton's approach, instantaneous tugs over great distances steer the planets in their orbits around the sun. If the sun suddenly disappeared, the "strings" would vanish, and planets would immediately begin to move in straight lines, following their own inertia. That change of behavior would happen even before the last rays of sunlight reached each planet, as light takes time to travel.

Thinking that such instant and remote action was unphysical, a bit like telepathy, Einstein constructed general relativity in a manner such that the crinkled fabric of spacetime serves as the go-between. The sun's massive presence creates a gravitational well by warping spacetime in its vicinity, a bit like when one steps into a bathtub and displaces the water. Such disturbances ripple away from the source, affecting the motion of other things. In the bathtub that might mean that rubber ducks, little boats, and other floating toys start to bob up and down. In the case of the solar system, the sun's gravitational disturbance radiates outward throughout spacetime at the speed of light, creating troughs that compel planets to move in curved orbits. The planets strive to move in straight lines, but the warped geometry in their regions forces them into curves.

After Einstein completed the general theory of relativity in 1915, he attempted to use it to model a universe that is static overall. Believing in ironclad determinism and eternal cosmic laws, he hoped that while mass might cause local perturbances, the global state of the cosmos would remain unaltered over time. In other words, though the stars might move in the sky, their concerted behavior, taken on average, would represent a universe as unchanging as a slab of granite. The permanence would not be preordained, as in Newton's construct, but rather a natural physical consequence of the theory.

Yet, much to his disappointment, the equations Einstein composed rebelled against such rigidity. Their solutions depicted a universe that either expands or contracts over time. In physics, the solution of an equation is a mathematical description that matches correctly, like a key that turns a certain lock. Einstein tried to find the perfect fit with a static universe but could do so only by tampering with his original equations—something like calling in a locksmith to change the locks to accommodate a beloved old key. The addition Einstein made, called the "cosmological constant term," was an extra "fudge factor" specifically included to counteract the destabilizing effects of gravity and yield the answer he expected. Indeed it did produce static solutions, but at the cost of making the theory somewhat more complex. Moreover, astronomer Edwin Hubble's 1929 discovery (aided by the work of other astronomers such as Vesto Slipher) that all distant galaxies are moving away from each other (and from us) would show that space does indeed grow over time. This led Einstein to remove the extra term and concede that the cosmos is expanding. Therefore he never realized his goal of vindicating Mach's ideas about inertia.

Wheeler discussed with Feynman, in the face of such developments, whether Mach's principle still had meaning, and, if so, what its physical basis was. He loved bringing up with Feynman (and others) abstruse philosophical questions and thinking of ways to test them. Feynman disliked abstractions but relished the testing part. That's one reason they were well matched.

As physicist Charles Misner, who would study under Wheeler in the 1950s, noted, "Wheeler was very much influenced by Niels Bohr, whom he treated as his second mentor. Bohr was very definitely in the European school of thinking: [emphasizing the] philosophical as well as technical [aspects of physics]. Most American physicists [such as Feynman] thought all the arguments about [the abstract, philosophical] interpretation of quantum mechanics were irrelevant to what they were doing."

PARTICLE PING-PONG

Human dialogue is like a game of table tennis. A typical interplay might include a conveyance of ideas, an exchange of jokes, banter about personal issues, or countless other modes of conversation. One player serves, and the other returns, like in a ping-pong match. Then the first responds to the second's play, the second rebounds, and so forth until the topic is exhausted. Wheeler and Feynman became adept at fitting their game to the mood of the day, switching in a pinch from witticisms to insights and lobbing these back and forth until the time came to move on to another kind of volley.

Elementary particles similarly interact with each other in pairs through a kind of exchange. Unlike human dealings, however, particle interactions have only a few fundamental kinds. Today we know that nature offers four basic types: gravitation, electromagnetism, and the strong and weak nuclear forces. At the time of Feynman's graduate studies, the latter two—referring to ways nuclei might bind or decay—were little understood; he would help decipher them later in life. Physicists didn't even know if these were separate interactions or the same force. Rather, they spoke of the "meson force" as the means by which protons and neutrons—the nuclear particles, or "nucleons"—cluster together by exchanging mesons. (Today we know that other particles, called "gluons," do the sticking together and that yet other particles, called W^+, W^-, and Z^0, convey the decay-inducing weak force.) Wheeler had spent much of his time with Bohr trying to understand why nuclei sometimes bind tightly and other times split apart. Their models worked empirically but still weren't complete.

Wheeler had a roving mind and an active imagination. He emitted one idea after another like a blazing furnace stoked with atomic power. Sticking to one topic would never do for him. He didn't even want to limit himself to the study of one fundamental force. Throughout his life his interests would pivot between studies of the nuclear, electromagnetic, and gravitational interactions.

In a different time, the idea of developing a unified theory of all the forces might have attracted Wheeler. However, he saw how Einstein, over at the neighboring Institute for Advanced Study, bashed his head against the wall, again and again, with just such attempts and generally met with ridicule for doing so. Einstein hoped unwaveringly that he could somehow expand general relativity into a theory of everything—describing all the forces geometrically while eliminating the need for probabilistic quantum theory.

Wheeler and Einstein lived in the same neighborhood, briefly shared the second floor in Fine Hall before the institute moved to its own space, and knew each other well. Striving in vain for such unification since the mid-1920s, Einstein had largely ignored modern developments, such as nuclear and particle physics, in his efforts. Because physicists, for the most part, viewed him as a relic, few ventured into the esoteric realm of gravitational theory—associated with success in the 1910s but with Einstein's failed quest since then.

The greatest achievement in gravitational theory of that period was largely ignored. A paper by University of California, Berkeley, physicist J. Robert Oppenheimer and his student Hartland Snyder, "Continued Gravitational Contraction," published on September 1, 1939, showed that a sufficiently massive star, after burning its nuclear fuel, would collapse into a compact object so dense and gravitationally powerful that not even light could escape. In the 1960s Wheeler would embrace this scenario, promote it using the term "black hole," and focus his attention on its bizarre implications. But in the 1930s his interests lay elsewhere.

Coincidently, on the same date the article appeared, Bohr and Wheeler published their influential "The Mechanism of Nuclear Fission," explaining why certain types of nuclei are easier to break apart, in the same prestigious journal, *Physical Review*; World War II began in Europe (as noted earlier); and the Wheeler family moved into a spanking new house at 95 Battle Road in Princeton—on property sold by the institute to construct a housing development. It was time for Wheeler to explore new theoretical vistas, and Feynman would be the perfect fellow voyager.

SCATTERED SHOWERS

Well before turning to the study of nuclear fission, Wheeler had developed a strong interest in scattering. Scattering happens when particles interact with each other and get deflected, as when a ball, hit with a racquet, bounces off in a seemingly haphazard direction. It occurs on both the classical (mundane) and quantum scales. Physics likes to make predictions, so in the case of a tennis maneuver, a clever theorist could use data about the ball's approach to the racquet to calculate its likely deflection. That is a classical problem, handled well by Newton's venerable laws of motion.

Wheeler was more interested in Compton scattering: a quantum process on the subatomic level not readily explained by Newtonian physics. Identified by American physicist Arthur Compton, who won the Nobel Prize for his discovery, the Compton effect involves light scattered by an electron. Shine light on an electron, and the electron acquires energy and momentum (mass times velocity), which hoist it in a certain direction like a hurled javelin. In the process it emits light of a longer wavelength (distance between peaks) than the original, aimed at an angle different from that of the electron. For visible light, wavelength corresponds to color, so the emitted light will have a different hue from the original, skewing toward the redder end of the spectrum. Normally, though, Compton scattering uses invisible

X-rays, in which case the emitted light will be X-rays of a longer wavelength.

The importance of the Compton effect is that quantum theory precisely predicts the difference between the initial and final wavelengths, along with the scattering angles of the electron and the emitted light. How it performs that feat reveals the essence of the quantum idea, first proposed by Max Planck in 1900 and refined by Einstein in 1905 in what is called the "photoelectric effect." The term "quantum," signifying "parcel," refers to the notion that light comes in packets of energy. These smallest quantities of light—waves apportioned into particles like boxed-up Slinkys—have come to be called "photons." As most of the light spectrum is invisible, aside from the narrow, optical range of colors from red to violet, the vast majority of nature's photons are invisible as well.

Photons serve as the exchange particles of the electromagnetic interaction. Every time a charged particle, such as an electron, attracts or repels another charged particle via electricity and/or magnetism, a photon is bounced between them. Without such an exchange, the charges would simply ignore each other, and there would be neither attraction nor repulsion. Therefore, if your prized refrigerator magnet clings tightly, you can thank photons (invisible rather than optical ones) in their role as electromagnetic force carriers.

As Planck and Einstein theorized, the amount of energy per photon depends on the frequency (rate of vibration) of the light with which it is associated. Frequency, in turn, is inversely proportional to wavelength (the greater the wavelength, the lower the frequency, and vice versa). Therefore, long wavelengths, such as those of radio waves, correspond to low frequencies and low energies. Short wavelengths, in contrast, such as those of X-rays, correspond to high frequencies and high energies. In Compton scattering, the electron gobbles energy and momentum from the incoming photon and spews out a weaker photon of longer wavelength. Researchers have measured the Compton wavelength shift countless times, and it always matches what they expect from the electron's energy gain.

Realizing Feynman's mathematical virtuosity—for instance, his uncanny knack for solving difficult integrals—and sharp intuition about physics, Wheeler proposed that they embark on a joint research study of quantum scattering processes. "Everything is scattering!"

Wheeler would proclaim as their rallying cry. The problems Wheeler wanted Feynman to investigate stemmed back to an international physics conference he attended, held in London and Cambridge in October 1934, where researchers discussed how gamma rays (the most energetic kind of photons) hitting lead produced a kind of "minishower" of scattered particles. Analysis of the scattering by-products, he realized, would help sharpen the quantum tool kit.

Wheeler had been the first to propose, back in 1937, an accounting method for tabulating the results of scattering, later dubbed the "S-matrix" (or scattering matrix) approach. It is akin to tallying the results of a game of darts according to how many hit each concentric ring, as well as the bull's-eye itself. For darts, such data could be used to figure out the strength and aim of the players. Similarly, in scattering processes, the S-matrix could be used to try to reconstruct the interactions that transpired. Physicists call such analyses based on collected data "phenomenological," in contrast to more abstract theoretical ruminations.

Wheeler and Feynman spent much time discussing the galaxy of questions connected with various types of scattering events. Under the master's guidance, Feynman came to know the S-matrix method very well. He also became adept with diagrams depicting how particles interact. After a brief investigation of gamma rays on lead, they decided to focus on how electrons and photons scurried like deflected pinballs through materials with intricate structures. That dialogue did not directly result in any publications but proved a prequel to even more fundamental questions about how electrons interact.

THE WHIRLIGIG

Experimental particle physics transpired in those days through two methods. One was to observe natural products of decay, such as particles produced by radioactive substances or cosmic rays raining down from space. For example, the positron—like an electron but positively charged—was first identified in a stream of cosmic rays.

The emerging alternative to natural methods was to find an artificial way of accelerating particles, smashing them into targets, and observing the residue. The granddaddy of this concept was the famous experiment designed by New Zealand physicist Ernest Rutherford, which bombarded gold foil with alpha particles (helium nuclei, as it turned out). The vast majority went right through the foil, but a tiny minority bounced back. By scattering at sharp angles, they revealed the gold atoms' compact, positively charged inner cores—gold nuclei. Before that discovery, physicists had presumed that atoms had uniform interiors, like the dense filling of a chocolate-covered cherry. The gold foil experiment demonstrated, to the contrary, that atoms were mostly empty space—the nuclei being minuscule parts of the whole. Instead of a filled confection, imagine a chocolate bonbon shell the size of a blimp, almost completely hollow, with nothing inside except a tiny cherry pit in the very center. That gives you an idea of the size of the nucleus, compared to the atom. The surprising results showed the value of understanding the scattering process. No wonder Wheeler emphasized its importance to Feynman.

In 1932, British researchers John Cockcroft and Ernest Walton, working under Rutherford at Cavendish Laboratory in Cambridge, England, constructed the first linear accelerator: a device that used electric boosts to speed up charged particle projectiles to the desired energies before hitting targets. Researchers found that aligning several boosts in a row, to create a longer linear accelerator, made such systems even more powerful. These were used to crack open nuclei and explore their properties—part of the experimental impetus behind the theoretical work by Bohr and Wheeler.

Another major breakthrough in accelerator design, developed around the same time as the Cockcroft-Walton mechanism, was American physicist Ernest Lawrence's cyclotron: a circular accelerator. Instead of consecutive linear boosts, a cyclotron uses the same booster multiple times. Magnets steer the subatomic bombardiers around and around, exposing them to the electric boost again and again, until they are powered up enough for release. They hurl toward their targets, smash into them, and generate valuable data through analysis of the collision debris. Far more compact than linear accelerators, cyclotrons proved increasingly popular throughout the late 1930s. Many of the top-ranked universities, including MIT and Princeton, had such machines.

Almost first thing on arriving at Palmer Laboratory, Feynman asked to see the Princeton cyclotron. The physics department sent him to the basement, where it was housed. He strolled through a cluttered storage area and finally found the device. It was certainly not what he had pictured.

Feynman expected Princeton's cyclotron to be far bigger and flashier than MIT's. He knew that it had proven more effective in getting publishable results. Yet, much to his surprise, he discovered the opposite. Princeton's particle-smashing device was a mess. As he recalled,

> The cyclotron was in the middle of the room. There were wires all over the place, hanging in the air, just strung up by somebody. There were water things—there had to be automatic water coolers, and little switches, so if the water stopped it would automatically go on, and there were some kind of pipes and you could see . . . water dripping. There was wax all over the place, hanging, where they were fixing leaks. The room was full of cans of film at crazy angles on tables. . . . I understood it immediately, because . . . it looked like my kid laboratory, where I had everything all over the place. . . . I loved it. I knew I was in the right place. . . . Fiddling is the answer. Experimenting is fiddling around. It's . . . completely inelegant, and that was the secret. So I loved Princeton right away.

Upon seeing the cyclotron, Feynman immediately realized why John Slater at MIT had advised him to complete his graduate work at Princeton. Princeton's particle physics lab had a makeshift quality that had proven far more suitable for getting results. In Feynman's view, physics should be done in a versatile way, rigging up different configurations, completing trial after trial, until an experiment yields conclusive, reproducible results. That usually involves a flexible setup. Feeling like a boy surrounded by an elaborate construction set, he was satisfied that he had made the right decision.

As an aspiring theorist—the direction he seemed to be heading in under Wheeler's guidance—Feynman did not expect to make use of the cyclotron to collect data. Still, that maze of pipes and wires drew him as though it were his own personal playground. Like Wheeler, even in the midst of abstract calculations, he dreamed of tinkering with the real stuff as he had as a child.

One day, around the time they were discussing Mach's principle, Wheeler and Feynman got into a heated discussion about X-shaped lawn sprinklers that spin around. Clearly those common garden gizmos worked on the basis of Newton's third law of action and reaction. From each of the four spigots, the water squirting out would trigger an equal strength reaction, known as "recoil," in the opposite direction. Thus four jets of water, gushing clockwise, would automatically produce four recoil forces, pushing counterclockwise, that would propel the whole contraption around and around repeatedly, like a whirling dervish. North, south, east, and west, the whole lawn would be soaked.

Time reversal would become an important theme in Wheeler and Feynman's collaboration. The opposite of squirting is sucking. Suppose the sprinkler valves ingested water instead of spewing it out. That would lead to recoil of a different sort. Would the combined reaction twirl the sprinkler around as well? That is, would the time-reversed action of gushing water lead to the time-reversed result: spinning in the opposite direction? Would it rotate in the same direction as the intake instead? Or would it just be a dud?

The two debated the question for some time, wavering back and forth on the projected result. Like a polished lawyer, Feynman thought of reasonable arguments for each of the possibilities, driving Wheeler a little nuts. Wheeler asked other Princeton faculty for their opinions, which were all over the map. Surely resolving such a garden-variety riddle shouldn't be rocket science.

Getting sick of hypotheticals, Feynman decided to put the matter to the test by building his own miniature contraption from glass pipes and rubber tubing. He set it up in the cyclotron room, where a huge container of water, called a carboy, would offer a bountiful supply. To generate the pressure needed to make the water suck inward instead of spurting outward, he hooked tubing to the cyclotron's compressed-air supply. Gradually he cranked up the air pressure, but little happened. Finally he went full throttle. Boom! The apparatus exploded. Shards of broken glass mixed with water sprayed over the cyclotron, requiring a time-consuming cleanup. The physics department reprimanded Feynman and banned him from the lab. (The correct solution to the reverse sprinkler problem has been the subject of considerable debate over the years. Under practical circumstances, due to various environmental factors such as fluid turbulence, there would be a pronounced difference between the two directions.)

EXPERIMENTS IN TIME

Feynman was perpetually curious, not just about the physical world but also about its connection to the realm of human experience. However, he had little tolerance for speculation based on pure reasoning, intuition, or feelings. Everything significant should be testable, he felt; otherwise why waste time with guesswork?

A mixture of antielitism and machismo preserved from his rather shy high school days may have partially driven his disdain for nonscientific erudition. He had been terrified others would see him as effeminate and effete—as a "sissy." Although he loved reading, he had dreaded being seen as bookish—as a science geek or nerd in today's parlance. His relative incompetence at competitive sports such as baseball made matters worse. Math contests did not offer the same macho street cred. He was relieved when he found a girlfriend, Arline Greenbaum—a sweet but assertive aspiring artist from Cedarhurst, New York—and could thereby prove that he was a "real guy." She called him "Rich" (to others he came to be called "Dick"); he nicknamed her "Putzie." They managed to keep up a long-distance romantic relationship throughout his time at MIT.

There, he had taken a philosophy class—the closest thing he could find to science that would fulfill a humanities requirement—and perceived it to be utter nonsense. He found the professor's mumbled remarks as meaningful as radio static. Feynman distracted himself during the boring lectures by using a miniature handheld drill to punch tiny holes in the sole of his shoe.

One day a classmate informed him that he needed to write an essay pertaining to the course theme: consciousness. Feynman vaguely recalled having heard the phrase "stream of consciousness" emerge from the professor's stream of gibberish. That struck a chord, reminding him of a science fiction scenario that his dad had once mentioned about aliens who never sleep and wonder what slumber is all about. For his essay theme, he decided to experiment with how consciousness trails off when one goes to sleep. Twice a day, during afternoon naps and after going to bed, he would try to notice how his conscious awareness changed in the moments before dozing.

At some point in his self-monitoring, Feynman observed something truly remarkable. In the drowsy prelude to slumber, his consciousness appeared to bifurcate. Instead of a single stream, it became twin rivulets. In one rush of thoughts, he visualized cords wrapped around a cylinder and threaded through a set of pulleys—resembling some of the mechanics problems he had been grading for Wheeler. Feynman was a visual thinker, so that wasn't too surprising. Vividly picturing every detail, he started to worry that cords would get stuck and the device would jam. However, he also noticed a second thought stream in which he assured himself that the tension force would guarantee that the system ran smoothly. Curiously, in two parallel strands, he was simultaneously the anxious student and the reassuring teacher. Yet the twin perspectives somehow meshed together like the cords in the pulley system.

"Stream of consciousness," a term coined by psychologist William James, describes the illusion that thoughts seem to flow in a single current. Irish writer James Joyce and other early-twentieth-century notables such as T. S. Eliot and Gertrude Stein adopted it as a kind of literary style. Joyce's notoriously abstruse novels, such as *Ulysses* (1922) and *Finnegan's Wake* (1939), offer literary logbooks of the mind's ramblings. Joyce, in turn, influenced Argentine writer Jorge Luis Borges, who delivered in the early 1940s a startling cache of short stories (originally in Spanish but later translated) about chance, time, and the mind. Not that Feynman read or was swayed by such literature. Rather, his insights generally arose from his own deep thinking and experimentation.

After Feynman submitted his class essay, his growing understanding of his thought patterns led him to experiments with what is now called "lucid dreaming": trying to preserve a sense of awareness and control during dreams. Dreamtime can seem completely decoupled from ordinary time. In that strange nocturnal realm, time's steady forward progression no longer seems to apply. A popular book of the era, *An Experiment with Time* by J. W. Dunne, envisioned a kind of time travel in dreams. Feynman's own delvings astonished him with regard to how much he could tell his dreams what he wanted them to do.

Feynman's mind experiments continued at Princeton, turning more explicitly to the subject of time awareness. He had heard about a prominent psychologist's theory that chemical processes in the brain involving iron metabolism governed how time is perceived.

Feynman thought otherwise and decided to investigate what factors influence time perception.

Could it have something to do with heart rate, he wondered? Running up and down the stairs of the Graduate College and racing along its hallways, he counted the seconds to himself. His dormmates had no idea what drove his wild sprints around the building. Breathless, he couldn't tell them until later, when they were in the dining hall together. There wasn't much to say anyway, as the running made little difference to his time sense.

THE HYPNOTIST

Wheeler's role in all this was mainly chuckling at Feynman's stories. On a few occasions, however, his spirited student invited him to the Graduate College for events, where he witnessed Feynman's offbeat inquisitiveness firsthand.

Entrance to Palmer Physical Laboratory (now Frist Center), bookended by statues of Benjamin Franklin and Joseph Henry, Princeton University (*Source:* Photo by Paul Halpern).

For instance, one day a hypnotist came to campus to entertain a large group of graduate students. Feynman asked Wheeler to come along as his guest. Much to Wheeler's surprise, when the hypnotist asked for volunteers, Feynman marched right onstage. A few commands later, he was deep in trance. The hypnotist solemnly ordered him to walk to the other side of the room, pick up a book, balance it on his head, and return. Like a programmed robot, he complied without question. The audience was in stitches.

Skeptical of hypnotism, Wheeler hypothesized that Feynman was simply acting. But Feynman wasn't inclined to perform for others (unless in an actual theatrical production). Rather, Feynman asserted he genuinely felt compelled to follow the commands. The brain, he realized, might not always tell the truth and could, for instance, lead one to think that following certain commands was mandatory. Through perpetual self-analysis and experimentation, Feynman derived a savvy grasp of psychology. Arguably, his investigation of altered states of perception would help prepare him to delve into a quantum reality that blends multiple timelines. Due to the mind's prejudices and limitations, things are not always what they seem.

On Saturday evenings the Graduate College sometimes hosted a dance. When Feynman was lucky, Arline would take a break from her art school studies and a part-time job teaching piano to come down for the weekend. By that time, they had started to talk about marriage and considered themselves engaged.

Arline's warm embrace, loving smile, and buoyant optimism offered Feynman cheerful respite from his classwork and calculations. She encouraged his artistic, expressive side and thus lent him needed balance. Don't guide your life according to others, she urged. Be yourself!

Thanks in part to her influence, later in life he took up creative hobbies such as drawing sketches and playing the bongos. Drawn to the rhythm of drumming, he eventually became an aficionado of the musical styles of various African and Latin American nations. More so than anyone else in his life, save perhaps his parents, Arline left an indelible imprint.

During those Princeton dance nights, Arline often lodged with the Wheelers: John, his wife, Janette, and their two young children, Letitia, nicknamed "Tita," and James, nicknamed "Jamie." Their newly built house on Battle Road was only blocks away from the Graduate College. John and Janette had been married since 1935, when the two had lived in North Carolina. Letitia had been born in 1936 and Jamie in 1939 (before Feynman arrived). They would later have a third child, Alison, born in 1942.

Janette was very fond of Arline, seeing her as a strong-minded, independent young woman. Someone as hardheaded as Feynman needed that balance. The young couple's growing love reminded John and Janette in many ways of their own deep affection. Worried, though, that Arline was taking on too much and working way too hard, they offered her a relaxing break by taking care of her at their house. Grateful for their hospitality, she gave them several watercolors that she had painted.

SOUPY TALES

Even when busy with computation, Feynman never wanted to spend all his time in the solitary confinement of an office, library, or laboratory. Rather he found it healthy to interact with others, especially when his mental gears were momentarily stuck. He tried not to take theoretical physics so seriously that the rest of life passed him by. Science should be a joy, not drudgery. People were far more important than equations.

Taking after his father, Feynman loved the wide-eyed look of children when he introduced them to fun, baffling aspects of science. At his childhood home in Queens, he had enjoyed pointing out scientific curiosities to his kid sister, Joan, who was almost nine years his junior. (Richard had also had a baby brother, Henry, who died of a childhood illness in February 1924 at only four weeks of age—a devastating tragedy for the Feynman family.)

As a little girl, Joan had assisted Richard with his electronics experiments, earning a "wage" of four cents a week. A request for a glass of water became a lesson in circular motion when he spun it in front of her eyes and "miraculously" it didn't spill—until he dropped it. He pointed out to her the fairylike green glow of the Northern Lights and strongly encouraged her interest in astronomy—which eventually led her to a successful academic career in the subject.

While he was at Princeton, they continued to write to each other about the wonders of the night sky.

Despite Joan's budding interest in science, Feynman never tried to explain his research with Wheeler to her. Perhaps he thought it too technical—or too far afield from astronomy. Nor did he ever introduce Joan to his Princeton mentor, even when she was older. As Joan recalled, "I had no contact with Wheeler at all and Feynman didn't discuss his work with me."

During his many visits to the Wheelers' house, Feynman came to know their children well. He enjoyed amazing them with his science tricks. It was part of an image Feynman would later hone as a kind of "science magician" who astounded others and defied them to find explanations.

Letitia and Jamie remembered Feynman coming over to the house when they were very little and conducting an entertaining experiment. Feynman snatched a can of soup straight off the counter where Janette was preparing dinner and, as Jamie recalled, said, "I have a problem for you. You have two cans of soup that are identical, but one is frozen. The question is, if you put them side by side on an incline and let them go simultaneously, which one reaches the bottom first?"

Although he didn't verbally reveal the answer to the kids, Feynman based his science trick on the fact that liquids have different dynamics from solids. Solid contents, such as frozen soup, spin in tandem with their containers and therefore expend rotational energy, which draws energy away from their motion through space. Fluids, on the other hand, such as liquid soup, don't spin with their containers and are free to expend most of their energy on moving from place to place. That enables the liquid-filled cans to go faster. Therefore even without opening the can or shaking it, you can tell if its contents are fluid or firm.

After posing his challenge about how to guess the state of a can's contents, Feynman tossed the soup can in the air. He found another can that contained a solid, tossed that one too, and asked the kids to note which fell more quickly. Based on their observation, they guessed which held the liquid. He opened the can, poured out the soup, and showed them how delectable thinking about physics really is.

Along with the soup can incident, Letitia recalled another of Feynman's visits, when his casual attitude clashed with Janette's more traditional views about how a young man should behave. When Janette walked up to him as he was slumped in a chair, she found it impolite that he didn't rise to greet her. "I have an image of Feynman," Letitia said. "I have a feeling that my mother was talking to him and said that he should stand up when a lady is talking to him."

Hosting graduate students and other young scholars in one's home was fairly common in those days, especially for professors familiar with the European tradition of private houses doubling as research centers. For example, Niels Bohr and his wife, Margarethe, warmly welcomed young researchers into their Copenhagen home, interlacing cozy discussions with legendary Danish hospitality.

The Wheelers repaid the favor by hosting the Bohrs on several occasions. It was exciting for the children to have such a famous physicist and his wife at their house. Letitia fondly remembered meeting Mrs. Bohr. Alison had memories of those visits as well. As she recalled, "Niels Bohr sat in my mother's favorite red velvet club chair. He spoke very softly and it was hard to understand a word he said."

CHAIN REACTION

Despite Bohr's soft-spokenness, his admonitions held considerable sway in the physics community. His quiet remarks at a young researcher's seminar, depending on their tone, might bolster or stymie the speaker's career. When he seemed agitated, such as in his announcement of the German fission discovery, his fellow physicists certainly took notice.

After several physicists sounded the alarm about the possibility of Nazi weapons development, the immediate answer was silence. Washington can move very slowly. Although Fermi contacted the Navy Department in March 1939 and Einstein first wrote to Roosevelt in August of the same year, the president saw little urgency. Prodded again by Szilard, Einstein sent two more letters in 1940. That year, the US government allocated about $6,000 for nuclear fission research (about $100,000 in today's dollars, adjusted for inflation). Only on December 6, 1941, the day before the Japanese bombed Pearl Harbor and the United States entered the war, did the American atomic program, later code-named the Manhattan Project, begin in earnest with much greater funding.

Bohr and Wheeler's paper had revealed two possible fissionable materials to fuel a chain reaction: uranium-235 and plutonium-239. Generating each in sufficient quantities would require enormous technological advances. Uranium-235, constituting only a tiny fraction of uranium ore, had to be separated from the far more abundant uranium-238. Research had shown that chemical processes and other common methods to distinguish ingredients simply wouldn't work. Plutonium-239 presented a wholly different challenge. As an artificial element, it would need to be created in a nuclear reactor through the transmutation of uranium.

Many more hurdles loomed, such as determining the critical mass of the fuel needed to create a chain reaction, assembling and housing that material, and so forth. The Manhattan Project would prove an unequalled scientific and technological feat, calling into service many of the sharpest minds in the United States (and its close allies Canada and the United Kingdom). In separate roles and locations, Wheeler and Feynman would each be recruited.

Wheeler would later wonder if the Allies shouldn't have put a greater rush on the atomic bomb program. After all, more than two years elapsed between Einstein's first letter to Roosevelt and the project's start, and almost four more years had passed before the bombs were constructed, tested, and dropped. While many of his colleagues would regret the devastation caused by nuclear weapons, Wheeler pondered alternative history scenarios in which the Allies thwarted the Nazis much earlier. Might a speedier development and use of atomic warfare, he wondered, have saved millions of lives?

While the war was still an ocean away, however, Wheeler spent 1940 and 1941 deeply engrossed in theoretical projects with Feynman. At that point he considered the conflict a European problem and preferred to wrestle with ideas in tandem with his brilliant young protégé. Rather than thinking about the logistics of nuclear fission, they were studying how particles interact on a fundamental level. Feynman chose Wheeler as his official PhD advisor, and Wheeler gladly accepted, formalizing their close working relationship. Meeting at Fine Hall, Palmer Laboratory, and Wheeler's house, calling each other on the phone, and finding every way to electrify each other's imaginations, they began to lay the groundwork for a revolution in fundamental physics. War was ephemeral; scientific truths, eternal.

CHAPTER TWO

THE ONLY PARTICLE IN THE UNIVERSE

> Feynman, I know why all the electrons have the same charge and mass. . . . They're all the same electron.
>
> —John A. Wheeler, as reported by Richard P. Feynman

> His brilliance, the wildness of his ideas, apparently impossible ideas, did fall on fertile soil, because I never objected to what other people would immediately have objected to.
>
> —Richard P. Feynman, describing his working relationship with John A. Wheeler

The briny Atlantic crashes again and again into Rockaway Beach, beating out a ceaseless rhythm. Surf and sand undulate to the same beat as they did when Feynman was a boy. Hundreds of miles north, in rocky Maine, the pounding surf repeatedly hits High Island, where the Wheelers once vacationed. Great physicists have come and gone, but the swell of ocean waves has lapped the shores again and again since time immemorial.

Electric lighthouses dot the coastline, each beaming a luminous cone upon the otherwise murky maritime darkness. Just as jiggling water molecules make the ocean waves, jiggling electrons create the light waves. In each case, particle disturbances generate a cascade of oscillations, propagating through space. But there the similarity ends.

While water waves are a mechanical phenomenon that requires a material medium, electromagnetic radiation, which includes visible light, can travel through empty space as well as materials. The standard explanation is that electromagnetic waves form a duo of electric and magnetic fields oscillating perpendicularly to each other at the speed of light.

A "field" is a kind of landscape of a chosen property (e.g., electrical strength) depicting how that property's values vary from point to point throughout all of space. It is like a map with data (e.g., GPS coordinates, elevation, population density) associated with every single location. Fields that have single values at each point are called "scalar fields." Fields with multiple values at each point, detailing how the magnitude (amount) and direction of the chosen property change from location to location, are called "vector fields."

Think of a weather map to understand the difference between scalar and vector fields. At any given time, each point has a single temperature. Therefore, the temperature readings for each location form a scalar field. Each point also has a certain wind velocity, which, unlike temperature, has both a magnitude (speed) and a direction. Consequently, the map of wind velocities constitutes a vector field.

In classical electromagnetic theory, vector fields convey both electric and magnetic forces. These fields fill space like a boundless sea of energy. Electric fields represent the amount and direction of electric force per unit charge at each point in space. Magnetic fields characterize the amount of magnetic force per unit *moving* charge at each point (classically, only moving charges experience magnetism).

Fields not only act on charges; they are also created by charges. An electric charge or set of charges automatically generates an electric field. If the charge or set of charges happens to be moving, a magnetic field is created as well. The directions of the electric and magnetic fields created by a set of moving charges are, in general, at right angles to each other.

Equations developed by James Clerk Maxwell show how the motion of these fields propagates via a kind of domino effect. A changing electric field naturally creates a changing magnetic field perpendicular to it. The changing magnetic field induces a changing electric field, and so forth, leading automatically to a train of vibrations chugging through space. This can happen in a vacuum as well as through a material substance.

Kick-starting that locomotion requires just an accelerating charge, such as an electron jostled up and down in an antenna. That movement leads to an electric field wiggling up and down, along with a magnetic field jiggling back and forth (at right angles to the electric field). Those wiggles and jiggles trigger even more wiggles and jiggles, forming an electromagnetic wave. That wave travels at light speed through a pure vacuum and somewhat slower through material media.

If the wave bashes into another antenna, it frees any loosely held electrons within to move up and down too. In that manner, the pattern from the transmitting antenna might be reproduced in the receiving antenna. Radio signals are broadcast through such replication—patterns generated in a radio station might thereby be transmitted to a car radio.

In the case of lighthouses, the transmissions—produced in filaments rather than antennas—have shorter wavelength and higher frequency, placing them in the visible range. They blast out white light, or something close to it, detected by the watchful eyes of nocturnal sailors, who appreciate the distinction.

Today the idea of electromagnetic fields—pulsating through space in the form of electromagnetic waves—is almost universally accepted. This Maxwellian idea has been successfully modified to match the predictions of quantum theory. However, in the early 1940s, when John Wheeler and Richard Feynman were conducting their collaborative research, it wasn't yet clear how to develop a complete quantum picture of electromagnetism. For that reason, they would see no compulsion to include the field idea in their models. Instead they would consider alternatives that revisited the older Newtonian idea of "action at a distance": remote interactions between particles.

AN ELECTRON'S QUANTUM LEAP

Wheeler and Feynman were well aware of the triumphs and shortcomings of quantum mechanics. They knew that in some types of measurements, it offered splendid predictions, while in others it fell short. The failings included quantum calculations that blew up to infinity, like dividing by zero on a modern calculator and getting the answer "undefined." In their research together, they decided to tackle such defects. To set quantum physics on firmer ground, they opted to pick and choose from its established ingredients, judging which were absolutely necessary and which could be modified or even discarded.

To understand the transformations our protagonists decided to enact, let's step back to the early history of quantum mechanics. We'll look at both nonrelativistic (low-speed) and relativistic (near-light-speed) versions. Then we'll see which quantum elements our intrepid researchers kept and which they decided to change or eliminate in their joint efforts to reform quantum physics.

In 1905, Albert Einstein's theory of the photoelectric effect showed that the wave picture of electromagnetism couldn't be the full story. Conveyed via photon "wave packets," light comprises a hybrid of wave- and particle-like properties. Compton scattering, where photons carry energies and momenta (particle features) related to their frequencies and wavelengths (wave features) offers an excellent example of this.

Much of Niels Bohr's early fame derived from his model of the atom as something like the solar system, with electron "planets" circling a nuclear "sun." Instead of a continuous range of possible orbits, however, Bohr derived rules for a discrete pattern of rings, each with its own characteristic energy level. The model renders the electron energy levels as something like a round stadium with fixed, circular rows of seats. Just as at a concert with assigned seating you must stay in your row unless your ticket allows you to switch, electrons must stay in their levels unless they have a quantum "ticket" that permits them to move closer to or farther from the nucleus. To venture inward or outward, they must emit or absorb a photon, respectively. Each photon glows with a frequency associated with the particular energy exchange transpiring. Amazingly, the frequencies predicted by Bohr's model for hydrogen matched beautifully with the rainbow of colors seen in its spectrum—a triumph for the theory.

Bohr couldn't adequately explain the reasons for the quantum rules confining electrons to certain orbits when they weren't jumping. The conditions just seemed ad hoc. To help remedy that situation,

Louis de Broglie introduced the notion of matter waves. Drawing on the work of Einstein and Bohr, he proposed that electrons and all material bodies have wave- as well as particle-like properties. Like photons, they ripple but are confined in space and have wavelengths related to their momenta. His bold idea immediately put the constituents of matter, such as electrons, and the carriers of force, such as photons, on almost equal footing. Almost, but with an important distinction.

The key difference between the stuff of matter, called "fermions" (after a quantum statistical method developed by Enrico Fermi and Paul Dirac), and the essence of force, called "bosons" (after the statistics of Indian physicist Satyendra Bose, along with Einstein), has to do with a quantum factor called "spin." Spin is a bit of a misnomer because, unlike the crazy sprinkler rigged up in the cyclotron room, it has nothing to do with actual rotation. Rather, it pertains to a particle's sociability with others of its type. Fermions are resolutely antisocial beings; each has its own quantum state. Austrian theorist Wolfgang Pauli proved this rule, called the "exclusion principle." Bosons, in contrast, are outgoing enough to share quantum states.

If we think of quantum states as the seats of a minivan and ask how many particles could fit into the backseat, the answer for fermions is "one"; for bosons it's "as many as they like." Unlike fermions, two or more bosons might have identical quantum numbers (a set of parameters specifying the precise quantum state). If a taxi driver picks up two fermions at once, there had better be at least two seats—one for each, no sharing. Otherwise they might have to take separate cabs. Bosons, on the other hand, love to pack together into the same quantum configuration. If imagined as taxicab passengers, with their proclivity toward sharing they'd never have to wait long for rides.

Suppose you try to force two electrons into the lowest energy level of an atom—that is, the one closest to the nucleus. As fermions, they absolutely can't be in the same quantum state and thereby need to distinguish themselves. One of them occupies a spin state called "spin up," and the other must be in the opposite configuration, "spin down." The nomenclature pertains to a property called the "Zeeman effect," where the atom is placed in a magnetic field. The spin-up electron aligns with the field, and the other counteraligns, splitting their energy levels into slightly different values.

Originally, the proposers of spin, Dutch physicists George Uhlenbeck and Samuel Goudsmit, called it such because they thought the electrons were actually like charged spinning tops. Their magnetic reaction came from the direction of their supposed rotation, counterclockwise having an axis pointed up and clockwise directed down. When that turned out to be physically impossible—the rotations would need to be faster than the speed of light—the name stuck nonetheless. Hence, physicists continue to use a spin quantum number that has nothing to do with actual rotation.

In the mid-1920s, German physicist Werner Heisenberg and Austrian physicist Erwin Schrödinger developed competing proposals for explaining atomic properties in a more satisfactory manner than the Bohr model. Heisenberg's scheme was more abstract, using the mathematical tables called "matrices" to show the odds of changing from one level to another. Schrödinger's method, much easier to envision, consisted of an equation showing how de Broglie's matter waves would take shape in a particular region, given its energy profile—like jelly adapting to a mold. Each fit experimental data well, leading German theorist Max Born to suggest combining the two theories into a single explanation.

In Born's combined approach, the solutions of Schrödinger's wave equation are probability waves, called "wave functions," rather than actual globs of matter. Probability waves map out the chances of particles being in any given location, rather than where they actually are (technically, one must square the wave function to get the probability). They are like the bell-shaped curves depicting the odds of certain sums when rolling dice. The peaks and troughs of wave functions show, respectively, where electrons will more or less likely be found.

Wave functions are far from permanent. Sometimes, due to environmental factors, they gradually evolve. An example is an electron subject to a slowly changing magnetic field; its wave function evolves correspondingly. Other times wave functions suddenly transform from one configuration to another. As in Heisenberg's matrix mechanics, these abrupt transitions are not perfectly predictable but rather have certain odds, like flipping a coin or spinning a roulette wheel.

Schrödinger's equation, though useful and elegant, omits several key properties of electrons. It does not take into account their spin;

nor does it address the effects of Einstein's special theory of relativity, proposed in 1905, the same "miracle year" in which he developed the photoelectric effect. While general relativity applies to gravitation, special relativity, its antecedent, is especially applicable to particles moving at high but constant speeds. When we are grappling with energetic electrons, ignoring Einstein's monumental breakthrough simply won't do.

RELATIVELY SPEAKING

Einstein's motivation to develop his special theory of relativity stemmed from a perplexing contradiction between classical mechanics and electromagnetic theory, connected with the constancy of the speed of light. In his youth, he envisioned a thought experiment in which a runner tried to catch up with a light wave. If the runner somehow were fast enough, Newtonian classical mechanics would allow him to keep pace with the wave; Maxwellian electromagnetic theory, however, would ban him from doing so because in Maxwell's theory light's velocity appears the same for all observers, even those moving at an extraordinary speed. As though chasing an ever-retreating desert mirage, the runner simply couldn't catch up to the wave.

Einstein's resolution of the conundrum, special relativity, holds that measurements of space and time depend on the relative speeds of observers. A swift runner and a rigid onlooker might record different values for the distance traveled by a light beam and for the duration of its journey. From the vantage of the fixed observer, relative to the moving one, space might be compressed and time dilated. However, on dividing distance by time to get velocity, each observer would measure the same speed of light. Thus, light's speed, not recordings by rulers and clocks, serves as the universal standard.

Soon after Einstein proposed his theory, mathematician Hermann Minkowski found that placing space and time on near-equal footing most elegantly expressed it. Minkowski devised the concept of spacetime to represent the amalgamation of the two. It would prove a natural means of describing general as well as special relativity.

In Minkowski's vision, space and time were not independent but rather dual aspects of spacetime. Spacetime replaced three-dimensional space and one-dimensional time with a four-dimensional unified entity. Minkowski dramatically announced his four-dimensional reformulation at a 1908 scientific meeting. Proclaiming "space by itself and time by itself are doomed to fade away into mere shadows," he showed how a fusion of the two into spacetime would offer an objective, invariant way of describing the universe.

In Minkowski's revolutionary approach, every event has four coordinates, melding its three-dimensional location with its time of occurrence. Nothing happens in space alone; it must have a time stamp too. Pure distances and durations are banished in favor of the spacetime interval: the separation between events in both space and time.

The shortest spacetime interval, called "lightlike" and with a value of zero, is light's straight-line path from one occurrence to another. It is like a twine that ties two events together with no slack. For example, if we were standing on the top floor of the Eiffel Tower and aimed a beam of light toward a boat on the river Seine below, the luminous ray would link two different spacetime occurrences with the greatest efficiency. The first event would have spacetime coordinates corresponding to the three spatial dimensions of the Eiffel Tower's location and the time of transmission; the second would have slightly different spatial coordinates and a minuscule time increase corresponding to the beam's arrival at the boat. Nothing could travel faster or straighter than that beam. Hence, lightlike intervals are the gold standard for communication and the most efficient way for causes to produce effects.

We might equally well point our beam of light in another direction, such as toward a different vessel. Indeed, the range of possibilities is vast. By plotting it on a "spacetime diagram," with time as one of the axes and the spatial dimensions as the others, one might imagine the vast sweep of angles from which light might emerge from a point and continue along a straight line. If the diagram has two spatial dimensions, along with time, the range of possibilities for light's path through spacetime looks something like the sweep of a lighthouse beacon as it rotates or the shell of an ice cream cone. Accordingly, scientists call the fanned-out array of options a "light cone." As depicted in the spacetime diagram, anything traveling at the speed of light would follow the light cone. Typically, situated under the light cone is a second upside-down light cone showing how incoming beams might arrive. In other words, it depicts the range of luminous rays from the past. Together, the upright and upside-down cones form something like an hourglass, showing the past and future limits of light-speed travel.

The science of optics shows why light, when passing through a vacuum or uniform medium, travels in a straight line. According to the "principle of least time," proposed by mathematician Pierre de Fermat in the mid-seventeenth century, light always takes the quickest possible path through space. Because its speed is constant, to minimize its travel time, it must choose the shortest route. As any student of geometry knows, the shortest distance between two points is a straight line.

According to special relativity, anything with mass must move slower than light. Therefore massive things offer (in many cases, much) slower means of communication than light. We witness this during a thunderstorm when the lightning flash arrives sooner than the thunderclap. If we wait until we hear the sound—conveyed via air molecules that have mass—to take shelter, as opposed to doing so after seeing the flash, our decision will be somewhat delayed. Waiting until we see neighbors running to escape the storm will delay our decision even further. That's why light, which takes the least-time path, offers the ideal means of communication. Note that we include light in all its forms, including invisible radiation such as radio waves.

Plotted on a spacetime diagram, things moving slower than light must follow paths that lie within the interiors of light cones (where the "ice cream" would be). That's because within a given time interval, the slower-than-light objects couldn't cover as much spatial distance as light would be able to traverse. For example, the paths taken by sound waves would fall within the interiors of light cones, not on their surfaces.

Naturally, we count ourselves among such bodies moving slower than light. On a spacetime diagram our lives look like twisty pipe cleaners, snaking in space from point to point over time. We call such patterns "world lines." As we proceed from birth to childhood, through adulthood and old age to death, these twisty patterns intercept those of other people, offering webs of converging and diverging relationships. Upon death, humans' world lines end, but those of their constituent molecules carry on. On a subatomic level, such as in the case of protons (the positively charged nuclei of simple hydrogen atoms), world lines can span billions of years.

Suppose an intelligent being, advanced beyond our comprehension, somehow obtained access to the complete spacetime diagram of the universe. The world lines of everything that ever existed or will exist—past, present, and future—would be etched within such a cosmic "crystal ball." From the creature's perspective, time would seem as frozen as a block of ice. Nothing could ever change, because it would already be foreseen. Such a timeless vision is often called the "block universe."

Philosophically, Einstein came to accept such a timeless worldview. As he once wrote, "To us believing physicists, the distinction between past, present and future has only the significance of a stubborn illusion."

To bring any physical theory in line with special relativity requires replacing any references to space and time as independent parameters with a unified spacetime. Consider, for example, Schrödinger's equation. It shows how wave functions behave in space and develop over time. Therefore it does not conform to special relativity. An alien living in a block universe, who lacks the notion of change over time, would have no idea what it meant. Schrödinger had tried, but failed, to come up with a relativistic equation that predicted electron behavior. It took an English physicist, with whom he would share the Nobel Prize, to devise the correct equation.

SEA OF HOLES

Bristol-born physicist Paul Dirac was notoriously reticent. Ask him an involved question, no matter how convoluted, and he would more than likely simply answer yes or no. An industry of stories arose about his parsimony of speech and social awkwardness. One famous tale involved his wife, Manci, who happened to be the sister of Eugene Wigner. Reportedly, he once introduced her simply as "Wigner's sister," implying that he didn't know her otherwise.

Luckily, in constructing equations, brevity and simplicity are ideal. In the late 1920s, Dirac developed a precise new lexicon for quantum mechanics. His clear-cut notation for quantum states and transitions is still used today. Along with codifying nonrelativistic quantum mechanics, he set out to describe a relativistic version too that included electron spin. By 1928, a scant few years after standard quantum physics and the idea of spin had emerged, he was victorious.

On paper, the Dirac equation, as his relativistic description of electrons is called, is one of the briefest in physics; yet it contains worlds of implications. It casts electrons in terms of special wave functions called "spinors" that transform according to distinct mathematical rules. It weds not only space and time but also energy and momentum in accordance with special relativity. Thus a spinor doesn't truly evolve over time but rather persists in a timeless block universe.

Weirdly, though, Dirac noticed that for every negatively charged solution of his equation, there was a positively charged counterpart of the same mass. Thus it predicted something like electrons, only of opposite charge. Protons wouldn't do; they are far more massive than electrons.

Stumped to explain the extra solutions, Dirac contrived an innovative "nautical" hypothesis involving holes in an infinite sea of energy. The universe, he conjectured, is full of an energy fluid (a reservoir of filled electron states) from which electrons emerge from time to time. Whenever they leap from that ocean, they leave behind a hole of the same mass and opposite charge—like bubbles that arise when a submarine surfaces. Hence, electrons are always twinned with holes.

In 1932, experimentalist Carl Anderson found evidence supporting Dirac's hypothesis in traces of cosmic rays raining down upon Earth. By examining particle tracks within a device called a "cloud chamber," he discovered a new subatomic body with a mass identical to the electron and the same value of charge, only positive instead of negative. Positive and negative particles curve in opposite spirals in the presence of a magnet, which is how he could tell the difference.

The "positron," as Anderson called the new particle, matched Dirac's theory perfectly. In short order the physics community embraced the idea of antiparticles—oppositely charged counterparts to ordinary particles. Numerous experiments confirmed that positrons were just as real as electrons, albeit far less common in nature. The concept of "holes," however, fell by the wayside, as it was extraneous to the description.

Rarely has experimental explanation so quickly followed a hypothetical solution. The discovery of positrons opened the gates to a vast menagerie of antiparticles, including negatively charged antiprotons. Today scientists believe that matter and antimatter particles equally populated the early universe, but certain asymmetric interactions led to the present-day imbalance.

Dirac's theories won him considerable acclaim and a widespread reputation as a mathematical genius. Physics students of the 1930s came to know him well due to his popular textbook, *The Principles of Quantum Mechanics*, which laid out his systematic approach to the subject. More than any other treatise of the time, it showed how quantum mechanics was a logical, highly predictive subject that nevertheless contained major gaps—including calculations that blew up into impossibly infinite values. It dared young physicists to try to bridge such canyons.

FINDING WIGGLE ROOM

At MIT, Feynman had carefully devoured Dirac's textbook and embraced its challenges. In particular, he found its final chapter, "Quantum Electrodynamics," an intriguing puzzle. Dirac had meticulously derived expressions for how relativistic quantum mechanics applies to electromagnetic interactions between electrons. The math was impeccable; yet the results were impossible.

In calculating the total energy, Dirac found that he needed to add an infinite number of terms. That wasn't necessarily a pitfall; sometimes even an infinite sum converges to a finite number. But Dirac's tally diverged instead. It blew up to infinity—like adding one plus two plus three, and so forth, on a calculator and continuing indefinitely. Only by setting an arbitrary cutoff could he obtain a realistic, finite answer. The brilliant physicist could not find an unequivocal way out of this pickle.

Feynman examined Dirac's calculation carefully, seeking a better method. Let's take a look ourselves. As Dirac pointed out, if two electrons interact at light speed, the signal between them must follow the strand of a light cone. While such a causal connection ordinarily happens forward in time, mathematically, one might consider backward-in-time signals as well. In the language of Feynman's day, the future-directed signal was called "retarded" and the past-directed signal was called "advanced."

Because electromagnetic waves travel at light speed, the light cone emanating from one electron thereby represents the range of other electrons at various times that might interact with it. They are on its "radar." Conversely, if an electron at a certain time is not part of another electron's light cone, the two are not on each other's "radar" and cannot affect one another.

Imagine two interacting electrons as two tall rocking chairs, connected by a clothesline that represents a strand of the light cone. While the light cone is a purely mathematical entity representing the time delays and spatial distances associated with light speed, we are here using something more tangible to symbolize it: the clothesline simply shows how cause and effect are connected. Rock one chair back and forth, and a signal, sent via the clothesline, will reach the other a little bit later, causing it to rock back and forth too. For electrons, the timing would match the speed of light.

Yet, if Maxwellian electromagnetic theory is strictly right, this analogy lacks an important ingredient: electromagnetic waves, or in quantum parlance, photons. A strand of a light cone represents a delay, but it does not automatically include an electromagnetic transmission that follows that path. Logically, just because two events are separated in such a way that electromagnetic waves could travel between them doesn't necessarily mean that they actually will. Nonetheless, following the standard prescription of Maxwell—accepted then, as now, by the bulk of the physics community—Dirac included electromagnetic waves as the medium of the electrons' interaction. How else could they "talk" to each other, such that one "tells" the other what to do?

Dirac characterized the waves—pulsating electromagnetic fields—as an array of harmonic oscillators, essentially springs, of different frequencies (rates of vibration). Why springs? As the simplest representation of anything oscillating, they are well understood. Quantum mechanics predicts their energies as a multiple of their frequencies.

Much to his frustration, Dirac found an infinite array of modes of vibration (technically called "degrees of freedom"), leading to a divergent sum of energy contributions. Hence the calculated total energy was infinite, which physically clearly wasn't the case. He could only get a realistic answer by cutting off the calculation at an arbitrary point.

In our clothesline analogy, since it would be silly to hang springs, let's imagine adding an array of sheets flapping in various rhythms. We hang sheets one by one, each vibrating differently. Soon, however, we discover that there is an infinite array of possibilities. Nevertheless, we wish to be comprehensive and include every type of oscillation. In a frenzy, we keep adding more and more sheets, until we are beyond exhaustion. The clothesline carries a thicker and thicker collection that never seems to stop!

After reading Dirac's account, Feynman started to think that maybe the light cone strand itself was enough. What if there were no electromagnetic fields and just a pure causal connection between the electrons? The result would be delayed action at a distance. While traditional Newtonian action at a distance didn't have a time delay, spurring the development of field theories such as general relativity, it could readily be included. The electrons would communicate remotely with a time delay built in, modeled by the light cone. That would ensure that cause and effect transpired at the correct pace—the speed of light—even if nothing was actually traveling between the electrons.

Leaving out the fields, Feynman boldly thought, might well be the trick to avoid the infinite sum. The only signal would be a direct interaction between the electrons. Simply shake one electron, and the other would shake as well, like the rocking chairs connected by a clothesline without the burden of hanging sheets.

Reviving action at a distance—after Maxwell, Einstein, and others had long dismissed it—might seem folly. Leaving out the media—namely, intervening fields—that convey forces from one place to another might seem counterintuitive. Yet this was a time of extraordinary revolution in science. So many aspects of subatomic physics had seemed bizarre at first, such as electrons jumping suddenly from one atomic level to another.

Feynman believed that action at a distance, with delays included, was worth reconsidering—especially given the alternative of grappling with unacceptably infinite sums in calculations. Maybe in the quantum world—for tiny distances yet to be observed directly—Maxwell's laws needed modification. Believing only what he could prove himself, Feynman was open-minded enough to test radical alternatives, such as abandoning standard electromagnetic theory on the subatomic scale.

Another well-known fact about electrodynamics similarly steered Feynman to consider leaving out the electromagnetic fields. In calculations involving either classical or quantum electrodynamics (as known at that time), the electron seemed to have infinite self-energy. Self-energy is the energy needed to assemble a particle or other configuration from scratch—much like the resources required to construct a building, including materials and labor.

By standard definition, self-energy includes the particle's rest energy (related to its mass by Einstein's famous equivalence formula), along with its interactions with the electromagnetic field it generates. For a finite-sized particle, such a calculation is feasible, because the field strength diminishes with distance from the center. One might determine how much energy it takes to build a finite ball of charge, given the forces generated by that ball on itself via the fields it harbors. It is like assessing the impact of the roof of a building on the floors beneath (and supporting) it.

However, if one assumes that the electron is a point particle, with infinitesimal size, its field at that central location is infinitely strong. Therefore the energy term corresponding to the interaction between an electron and its own field is infinite as well. Calculating the self-energy of an electron thereby yields a value of infinity—clearly not a realistic physical result.

A simple remedy, Feynman pondered, would be to forbid electrons from interacting with their own fields. Instead, the fields would be gone. The electrons would interact directly with each other and not at all with themselves. Then the electron's self-energy would match the value derived from its mass, using Einstein's conversion formula pure and simple. It would be finite and reasonable.

FACING RESISTANCE

By the time Feynman was at Princeton and working with Wheeler, he had discovered a major problem with leaving the fields out of his delayed-action-at-a-distance hypothesis, developed at MIT. A well-known phenomenon called "radiation resistance" demonstrated that electrons and other charged particles are harder to accelerate than electrically neutral (chargeless) particles. Speeding up protons, for instance, would be harder than doing the same with neutrons, even though their masses were comparable. The logical explanation was that charged particles generated radiation—in the form of electromagnetic fields—that acted back on them and hindered their motion. In our rocking chair and clothesline analogy, that would be like finding the chair harder to push because of the hanging sheets deterring its motion. Neutral objects wouldn't carry such weight, explaining their relative mobility. To replicate the observed radiation resistance, were electromagnetic field interactions needed after all, wondered Feynman? Or could there be another way?

When Feynman's scattering research with Wheeler had reached a lull, he decided to mention the predicament and his frustration with resolving it. Feynman brought up one conceivable solution that had popped into his head. Suppose radiation resistance was a direct effect on an electron due to all the other charged particles in space instead of the electromagnetic field. Jiggle an electron, and all the other charged particles would react, sending signals back to the original, somehow conveyed remotely without the field. The compendium of reactions of the other charged particles would produce a force on the original electron, explaining why it was harder to accelerate.

Revising our analogy, any given chair would be attached to many others by numerous clotheslines. Rocking it would cause all the others to rock as well, tugging back on the first and hindering its own motion. No hanging sheets would be required to explain the resistance.

Listening very carefully, Wheeler immediately pointed out a few sticking points. If radiation resistance depended on how other charged particles affected an electron, their specific properties—mass, charge, distance, and so forth—would matter. Therefore, theoretically each electron would have a distinct amount of radiation resistance, depending on its specific environment. This is not observed in nature. Rather, every electron's radiation resistance, taking into account its motion through space, is the same.

Also signals would take time to travel from the electron to the other charged particles and back again. But experiments had shown that radiation resistance happened instantaneously, not with a delay. Finally, if you summed the reactions of all the other charges in the universe, they would add up to infinity. One mathematically impossible situation would replace the other.

Feynman was stunned by how quickly Wheeler identified his model's key weaknesses—almost as though his mentor had spent countless hours test-driving it and finding its rattles, clunks, and glitches. But Feynman had just presented it. He felt like a complete fool.

As a matter of fact, Wheeler had thought deeply for years about the idea of replacing the field approach to electromagnetism with the more direct concept of action at a distance. To simplify physics by reducing its components, he endorsed resuscitating the original Newtonian idea of forces as "invisible threads" connecting distant objects rather than having a physical intermediary. Michael Faraday and James Clerk Maxwell had developed the field idea to make electromagnetism local and tangible, but perhaps on a quantum level they were wrong.

Action at a distance, Wheeler thought, would make particle physics simpler by making electrons sole masters of their own fate. They would govern their own interactions without an intermediary. He pondered the idea of "everything as electrons," including not just electromagnetism but also the other particles and forces. This would lend a beautiful unity and simplicity to the universe.

Part of the motivation for resurrecting action at a distance in quantum electrodynamics stemmed from a growing understanding that many quantum phenomena coordinate their features remotely. Such remote interplay, called "entanglement," transpires when two particles with complementary values of a quantum number (a parameter designating the specific quantum state) such as spin are linked in the same system, no matter how distant they are physically.

Take, for example, a pair of electrons residing in the lowest energy level of a hydrogen atom. Pauli's exclusion principle mandates that they cannot have precisely the same quantum number. Consequently, they must be in opposite spin states; if one is spin-up, the other is spin-down. Until researchers measure the spin states, however, they do not know which is which. Therefore, before such a measurement each electron is in a superposition (quantum state mixture) of the two spin possibilities.

Now imagine that scientists widely separate such an electron duo before gauging the spin of either one. One could be rocketed to the moon while the other remained on Earth. Despite the enormous distance, if an astronaut on the moon recorded the spin of the lunar electron to be "up," the other would by necessity instantly revert to "down," and vice versa, in a kind of quantum seesaw.

Believing such instant coordination impossible—akin to pseudoscientific claims of psychic communication—Einstein called it "spooky action at a distance." How could one electron know in advance what an experimenter might do to another? A 1935 paper that he coauthored with Boris Podolsky and Nathan Rosen (but written largely by Podolsky) described the "EPR paradox" and highlighted supposed contradictions that would arise via entanglement, such as particles having to predict which of their properties was about to be measured.

Most quantum physicists ignored or dismissed Einstein's critiques. As Bohr—the community's "philosopher king," if there was one—freely acknowledged, the field embraced contradictory aspects, such as wave- and particle-like features. Bohr called this union of opposites "complementarity." As an emblem of this, he included the Taoist yin-yang symbol, with its swirling medley of darkness and light, in his family crest.

Philosophically, Wheeler began his career firmly in Bohr's camp, accepting quantum uncertainty and complementarity as fact. Yet he came to know Einstein very well and to appreciate his style of reasoning. Einstein lived only a few blocks from him. Sometimes Wheeler would see Einstein walking down the street with his two assistants at the time, Peter Bergmann and Valentine Bargmann. The three were attempting to develop a unified theory of the natural forces that they hoped would relegate the nonlocal, probabilistic aspects of quantum physics to the dustbin, replacing them with a local, deterministic extension of general relativity. Agreeing with Bohr that such efforts were misguided, Wheeler nevertheless admired Einstein's independence of thought. He hoped that developments in theoretical physics would eventually offer explanations so compelling that both Einstein and Bohr would embrace them.

Unlike Einstein, Wheeler didn't see action at a distance as taboo. Rather, in his view, entanglement clearly showed that quantum physics was nonlocal. Given electrons' remote coordination of spin states, Wheeler was willing to take the leap of recasting their electromagnetic interaction as nonlocal as well. Wiggle an electron, and another might wiggle too in a kind of long-distance conga line. The key difference would be that in the case of electromagnetism, there must be a time delay. Special relativity mandated that the conga line could proceed no faster than the speed of light.

ZIGZAGGING THROUGH TIME

Once Wheeler latched onto the radiation resistance problem for electrons, he and Feynman put their minds together to try to model the effect without electromagnetic fields. They needed to find a way that any electron accelerating at a given rate would experience the same resistance, without delay, no matter what the arrangement of other charges in the universe. It was like ensuring a car's brakes would work the same way each and every time, no matter the road conditions or the actions of other vehicles.

In fashioning a more realistic model, Wheeler imagined what would happen if an electron was speeding up and met with resistance due to its neighborhood of charged particles. The accelerating electron would first send out some kind of a signal. Then, like a mirror reflection, something in the environment would signal back. The second signal would impede the electron's motion. Because the effect happened instantly, there couldn't be a time delay between the first signal leaving and the second hitting back. The second must arrive exactly when the first one was emitted. Logically, Wheeler realized, that could only happen if the second signal journeyed back in time to that moment.

Wheeler knew that Maxwell's equations are fully symmetric in time. For every wave solution moving forward in time, another is moving backward in time. The latter, called the "advanced solution," is traditionally ignored because everyone knows that clocks run forward, not backward. However, Wheeler was extraordinarily open-minded and wanted to see what happened if the advanced solution was included. The instant the electron broadcast a signal to the medium (its environment, essentially the net effect of everything else in the universe) announcing its presence, the medium would fire back a signal that arrived just when the first left. For technical reasons, the medium needed to be a perfect absorber, picking up every signal. Therefore, in the time-reversed solution it would act as a perfect emitter, sending back a clean signal, unadulterated by any material effects of the medium. The result would be an immediate hindrance of the electron's acceleration, independent of other particles' properties.

In our clothesline analogy, that would be like attaching one end of the line to a rocking chair and the other to a wall (representing the medium). Rock the chair, and it will send out a pulse. Reflecting off the wall, the signal will bounce back, wriggle through the line in the reverse direction, hit the chair, and hamper its rocking. Now imagine that somehow the wall sends its reflected pulse backward in time to hit the chair the very instant it starts rocking: that would be the weird effect of an advanced signal.

Wheeler's suggestion intrigued Feynman. Setting out immediately to calculate the effects, he tried different combinations of the outgoing and reflected pulses to produce a net effect that would model radiation resistance. Soon he found the right mix: a fifty-fifty hybrid of forward- and backward-in-time signals, which was thereby completely symmetric in time. He could describe radiation resistance without including electromagnetic fields, thus avoiding the problem of diverging energy that had caused Dirac and others so much grief. With photons banished, light would become the direct interaction of electrons, pure and simple. The construct would become known as the "Wheeler-Feynman absorber theory."

THE INTERROGATION

With Feynman's calculation in hand, Wheeler was beaming. Finding revolutionary potential in their new approach, he informed Feynman that it was time to hold a seminar on the idea. Both knew that the project wasn't yet complete; they had used classical, not quantum, methods. Fully addressing the self-energy problem of the electron and other pressing issues would require a complete quantum electrodynamics, which no one had successfully come up with at that time. As in the classical case, prior attempts to quantize (render in quantum form) electrodynamics had produced mathematically unsavory infinite terms for self-energy and other quantities.

A quantum theory would mean replacing the exact deterministic mechanisms of the classical theory with a probabilistic description based on mathematical functions called "operators." A certain haziness and indeterminacy needed to be included, to reflect the murkiness of quantum reality on the subatomic scale. With naive optimism, Wheeler said that he would quickly handle the quantum part, while Feynman prepared the classical talk. He assured Feynman that he would later deliver the quantum sequel lecture once that research phase was finished.

Although Feynman was understandably anxious about delivering his first research lecture, Wheeler calmed him and advised that it would offer him valuable speaking experience. Once Feynman agreed, Wheeler asked Wigner, the lecture series coordinator, to list the event on the department's calendar.

A few days before the seminar, Feynman was strolling down a Fine Hall corridor when he ran into Wigner. Commending him on an excellent topic, Wigner mentioned some of the professors invited to the talk. John von Neumann—widely considered a genius and one of the world's foremost experts in quantum measurement theory—would be there. So would acclaimed astronomer Henry Norris Russell, famous for his stellar classification scheme, among other achievements. What's more, Pauli, on leave from Zurich and visiting the Institute for Advanced Study, would be sure to participate. Finally, Einstein, who almost never came to physics department seminars, was intrigued by the topic and also planned to come. After hearing about all the "monster minds" likely to attend his talk, Feynman was even more a bundle of nerves—all tangled up inside like the cyclotron's wiring.

Wheeler was reassuring. If any questions proved too tough, he could handle them. Feynman calmed down and methodically prepared notes for the talk.

When the seminar was about to begin, Feynman started to put equations up on the board. Suddenly a white-haired man in his sixties with a thick southern German accent interrupted him. "Hello, I'm coming to your seminar," said Einstein. "But first, where is the tea?"

Pointing to the refreshment table, Feynman breathed a sigh of relief that he could answer at least one of Einstein's questions. The seminar began and turned out not to be so bad after all. Immersed in the calculations, Feynman forgot that an audience was watching him. He was in kind of a relaxing trance, as during the time with the hypnotist.

A sharp remark by Pauli about the mathematical limitations of the theory—which he didn't think would add up as promised—abruptly snapped Feynman back to reality. The Viennese physicist was notoriously hardheaded and critical. He had a diabolical gift for finding the structural flaws in any theoretical edifice and reporting these in the coldest, most direct manner possible. Immediately he saw that the model was crawling with mathematical bugs. It couldn't support itself and certainly wouldn't be a sound base for a quantum theory. So he simply commented that it wouldn't work.

Later, Pauli would tell Feynman privately that Wheeler's wish to quantize the theory was a pipe dream. He scolded Wheeler for not being honest with his student about the mathematical challenges of quantization. Wheeler would never deliver the promised quantum sequel, he harshly predicted.

In contrast, Einstein's reaction to Feynman's seminar was friendly but neutral. By that point, he was so focused on developing a unified field theory and so removed from quantum physics that he had little to add. He simply drew attention to the fact that it would be hard to connect the Wheeler-Feynman absorber theory with general relativity. Doing so would be a requirement for incorporating it into a unified theory of electromagnetism and gravitation. However, unlike Pauli, he was reluctant to dismiss it and thought it seemed fine as it stood.

Einstein was more helpful when, shortly thereafter, Wheeler brought Feynman to Einstein's house. Since 1936, Einstein had been a widower, living with his sister, his stepdaughter, and his secretary, each of whom had learned not to make demands on his time. Although he relished spending many hours alone, immersed in his own thoughts, he also enjoyed deep discussions with others about the philosophy of physics—especially young people such as Wheeler and Feynman, who might be receptive to his unorthodox ideas.

Wheeler asked Einstein point-blank if the notion of time-reversed signals made sense. Einstein was sympathetic. Referencing a paper he had written with physicist Walter Ritz, he expressed the view that fundamental physics should run equally well forward and backward in time.

Greatly encouraged by Einstein's counsel, Wheeler decided to ignore Pauli's naysaying and start thinking about ways to quantize the theory. As he tried to proceed, though, he encountered more and more mud in the road. Soon he found himself stuck in a rut.

To make matters worse, Wheeler had sent notice to the annual meeting of the American Physical Society stating that he would deliver a talk about the quantum theory of action at a distance. He had no idea what he would say but thought that he could at least deliver a progress report. He invited Feynman, who listened eagerly to hear how Wheeler solved (or at least attempted to solve) the problem.

Feynman waited and waited as Wheeler sketched out the details of the classical theory but made no mention of the quantum, then abruptly switched to a completely different topic. Feynman stood up, raised his hand, and interrupted Wheeler. The talk had nothing to do with its title, he complained. There was no quantum theory, as of yet.

Feynman did not intend to be rude; he simply felt that where science was concerned, being honest was the only unequivocal way to make progress. Even a misconception about a title might lead to an incorrect understanding of what was already known.

Wheeler agreed with Feynman's assessment. As they left, he admitted to Feynman that the talk had been a big mistake. He didn't have the answer yet and shouldn't have implied that he did.

As usual, Pauli was unnervingly right. Wheeler realized that, once again, he would need to rely on Feynman's razor-sharp mind to scrape away the mathematical muck and enable the project to move ahead. Yet he was too embarrassed, perhaps, to say as much. So without actually telling Feynman that he saw no way forward, he watched his student toil independently and make far more progress. Feynman's brilliance at calculations perfectly complemented Wheeler's perceptive philosophical musings, which often wouldn't otherwise end in fully realized results. While Wheeler was like Leonardo da Vinci in his clever schemes (those that remained sketches), Feynman was like Michelangelo in his stunning creations. Quietly, Wheeler made way for the scientific sculptor to hone his craft.

ARTFUL INTERACTIONS

Feynman had begun to appreciate art and relished having an artist as a girlfriend. The world was so much more than lifeless equations. Drawings showed the plain truth of things. Though math was fun and full of puzzles, fundamentally it was just a drafter's tool to model underlying mechanisms when they weren't apparent. While nature's escritoire contained hidden drawers filled with calculations, its gleaming visage in the sunlight was far more wondrous and revealing.

Art is timeless. So is love. Young lovers hope that their romantic vision will persist forever. When one latches onto a beautiful present, the past and future might feel immaterial.

Unfortunately, harsh rains can wash away watercolor dreams. As much as he rooted for Arline to succeed as an artist, Richard had begun to realize that the path ahead wouldn't be easy. She worked herself to the bone but barely made a living. Moreover, she had started to develop troubling medical symptoms.

One time when he visited her, he noticed an unusual bump on her neck. She rubbed ointment on it, but over many weeks it persisted, and she developed a fever. Thinking it might be typhoid, her family doctor urged her to check into the hospital, where she was briefly quarantined. Tests for typhoid came up negative, and she was released.

Soon she developed more lumps in the area of her lymph nodes. Her fever returned, and she went back for more tests. As her physician tried to determine the cause, Feynman leafed through medical books in the Princeton library. Frantically he tried to find the cause of her distress. After carefully researching her symptoms, he concluded that it was likely Hodgkin's lymphoma, a serious type of cancer. If he was right, he worried that it would cut short their plan to spend many happy years together.

Trying to be as honest with Arline as possible, he shared his research with her, with the caveat that sometimes laypeople reading medical books arrive at false, alarming conclusions. She appreciated his honesty and urged him always to be truthful. When she mentioned Hodgkin's to her doctor, he seemed concerned and wrote that in her chart as a distinct possibility. A diagnosis would require even more tests.

During one round, Feynman came up from Princeton and accompanied her to the county hospital. After some of the results came in, her doctor somberly took Feynman aside and told him that Hodgkin's was the likely culprit. If she had it, she would probably live a few years at most. The doctor suggested that Feynman keep the news quiet. There was no reason, he said, to shatter Arline's fragile emotions with such a dreadful prognosis.

Having promised Arline always to be honest and knowing how resilient she was, Feynman wanted to tell her the truth. Fearing she would fall apart, family members urged him instead to paint her condition more rosily. At first, he reluctantly took their advice. After reassuring her for a time that she only had glandular fever (today, more commonly called mononucleosis, or "mono"), he finally broke down and leveled with her. As anticipated, she faced the devastating news with great courage.

Because she clearly couldn't support herself anymore and needed a lot of help, Feynman decided that the prudent thing would be to marry her. He knew, however, that if he did, he'd lose his scholarship and be forced out of the graduate program. He'd have to find a job in a private company such as Bell Labs to support the two of them. He also wanted to complete his project with Wheeler, which offered him a needed distraction from thinking about Arline's health.

Despite her illness, Arline remained steadfast in her optimism. She was madly in love with Richard and wanted to encourage him in every way possible. Every time he was disheartened, she cheered him on. Whenever he found success, she was elated. For example, when he told her that he expected his work with Wheeler eventually to be published, she wrote to him, "I'm awfully happy . . . that you're going to publish something—it gives me a very special thrill when your work is acknowledged for its value—I want you to continue and really give the world and science all you can."

Feynman soldiered on with his doctoral research. Continuing his study of the interplay between electrons, he found that he could map such interactions handily using spacetime diagrams, with the horizontal axis representing space and the vertical axis, time. Diagonal lines—two-dimensional depictions of light cones—represented interactions transpiring at the speed of light, either forward or backward in time.

In such pictures, backward signals seemed just as logical as forward signals. Feynman saw no need to worry or philosophize about the lack of causality. In his view, nothing in the Einsteinian universe dictated that cause must always precede effect. Just as left and right could be reversed, so might future and past. Clearly causality was real—humans experienced it every day—but it did not have a bearing on particle interactions. Furthermore, as Wheeler had pointed out, most of the backward signals in the universe might get canceled out by forward signals, so violations of causality might rarely, if ever, be noticed directly.

Following Dirac's methods, Feynman represented the signals in terms of combinations of harmonic oscillators of various frequencies (vibration rates) and amplitudes (peak values). Such simple vibrating systems, akin to springs, possessed clear physical and mathematical structures and thereby offered ideal components of more complicated patterns. However, he broke with Dirac's methods by framing the interaction between electrons as occurring directly, rather than via photons, and excluding the possibility that an electron could interact with itself.

As much as he loved the determinism of classical mechanics, Feynman knew that the quantum process would introduce some murkiness. Heisenberg's principle of uncertainty mandated that position and momentum could not be known exactly and simultaneously. Such fuzziness suggested that it would be hopeless even to draw diagrams. Heisenberg himself considered visualization unnecessary and often misleading. Nevertheless, Feynman persisted. As a visual thinker, he needed to work with sketches, not just abstractions.

Somewhat promising medical news—compared at least to the dire earlier prognosis—helped spur him forward. A biopsy of the swollen gland on Arline's neck revealed that she didn't in fact have Hodgkin's. Rather, she had contracted tuberculosis of the lymph nodes, a serious condition, but one with a greater life expectancy. There was no cure at the time for the so-called white plague, but for lucky patients, methods to ameliorate symptoms sometimes led to recovery. She'd still need a lot of treatment, particularly recovery time in sanatoriums, but might survive for many years. The reprieve from her imminent death offered Feynman the opportunity to complete his degree before getting married. Wedding visions helped galvanize him to finish his doctoral project as quickly as possible.

FOLLOW THE LIGHT

At that time—before Feynman would perfect his revolutionary method of summing quantum histories—the standard way to transform classical into quantum was to replace variables, such as position and momentum, with mathematical functions called "operators." Those typically involve instantaneous changes with respect to space and time—known, respectively, as space and time derivatives—in the wave functions representing particle states. The most important operator, called the "Hamiltonian," constitutes a combination of operators representing kinetic and potential energy. It applies derivatives and other mathematical operations to a particle's wave function to yield, under certain circumstances, its total energy value.

In calculus, a derivative tracks how something changes slightly during an infinitesimal interval of space or time. For example, if you record a child's growth with a chart, the derivative of the height curve will tell you how much he or she grows in each instant. Derivatives therefore require a measure of locality (things happening at a certain point in space and time) and continuity (things not jumping suddenly from one value to another).

The Schrödinger equation, constructed around the Hamiltonian operator, nicely shows how a wave function's transformation in space relates to its transformation in time. The equation involves derivatives taken at each point and moment that dictate what happens at the next point and moment. Consequently, as a locally defined procedure requiring continuity from point to point, the Schrödinger equation was incompatible with the action-at-a-distance formalism that Wheeler and Feynman developed.

Dirac's equation, also involving derivatives, didn't fare any better for Feynman and Wheeler's purposes. Instead of using separate space and time operators, it combined them into spacetime versions, and it replaced standard wave functions with more complex variants, spinors. Nevertheless, requiring locality in spacetime, it was similarly ill suited to the action-at-a-distance approach.

Feynman realized that to quantize the theory he would need to start virtually from scratch. He would have to develop a means to link events separated widely in spacetime. In the action-at-a-distance representation of electromagnetism, in its quantum as well as classical form, two electrons would be associated through their remote interactions rather than through anything physically passing between them. Anything with a Hamiltonian wouldn't do.

While leaving photons out of the theory, he knew that he needed to include a light-speed delay. As Einstein had established, information travels at light speed; there was no way around that. In a spacetime diagram, the points representing two interacting electrons had to reside on the same light cone. The electrons would signal each other, whether forward or backward in time, at the speed of light. The path of light, therefore, offered a magnificent clue as to how to proceed further.

From his readings about optics and mechanics dating back to high school, Feynman had long been familiar with Fermat's principle of least time. It nicely predicts how light behaves and shows how the quickest path for light to travel through a particular substance is a straight line: a light ray. It also reveals how when passing from one material to another, light bends at certain angles, following what is called the "law of refraction."

One can demonstrate how Fermat's principle works by envisioning light from a source journeying along every possible path to a certain destination. Each lightwave has a certain phase. The term "phase" in physics refers to the amount of lag in a wave cycle. If two waves have the same phase, their peaks and valleys line up perfectly. If they are 180 degrees out of phase, the peaks of one wave line up with the valleys of the other. If they are out of phase by another amount, the peaks and valleys neither align nor mesh, resembling something like a zipper whose teeth are out of synch.

Light waves taking virtually identical routes typically have little phase difference. If two light waves take different paths, on the other hand, the time delay will often lead to a significant phase difference. Hence, proximity of paths offers the easiest way to guarantee minimal phase difference.

The distinction between similar and different paths shows up in the process of interference: adding up waves to form a single, representative one. Waves with little or no phase difference exhibit constructive interference, meaning their peaks and valleys add up to form a larger wave. Waves with phase differences close to 180 degrees destructively interfere, indicating that their peaks and valleys cancel each other out, leading to a flatter wave. Two waves that take similar routes thereby constructively interfere. Those with very different routes, and thereby a range of phase differences, have a variety of interference patterns that are not generally constructive.

Now here's where Fermat's principle kicks in. Consider the collective interference of the waves representing all possible paths from the source to the destination. Those waves that happen to be along routes taking the least amount of time—approximately the same trajectories—are roughly in phase. Therefore they constructively interfere, resulting in a large amplitude (peak size) sum. The other routes tend, in contrast, to cancel each other out, leaving a flatter profile. Thus the path of least time offers the most prominent way for light to travel, seen visibly as a beam.

In classical mechanics, objects don't always take the least time and/or shortest path to travel from one point to another. You'd be surprised if you threw a basketball toward a hoop and it traced a diagonal line like a flashlight beam. Rather it would follow a certain curve: a parabola. Nevertheless, as it turns out, you could use another principle of physics, called "least action," to predict its path.

Action is a peculiar quantity consisting of units of energy multiplied by time. Unlike variables such as position and velocity that differ from point to point and moment to moment, it is defined for an entire route, from one event in space and time to another. It is related to another quantity, called the "Lagrangian," that comprises the difference between the kinetic energy (energy of motion) and potential energy (energy of position) for a certain object (or set of objects). Briefly, the action is the integral (adding up) of the Lagrangian's values for each point in time along a certain path.

When you throw a basketball into the air, for example, as it rises its kinetic energy transforms into potential energy and the Lagrangian gets smaller and smaller over time. When it falls back down toward the hoop, its potential energy transforms back into kinetic, and its Lagrangian increases. Multiply the Lagrangian value for each time by the infinitesimal time interval, add these up using integral calculus, and you get the action for that path.

According to the principle of least action, proposed by Irish mathematician William Hamilton, an object takes the route that optimizes (minimizes or maximizes) the action. Typically, that corresponds to the minimal action. Therefore, if you compute the action for every possible path that a basketball might take, the lowest value matches the genuine trajectory. Mathematically, then, by computing the actions for all possible paths and minimizing that value, the result is a set of relationships, called Lagrange's equations, that describe the body's actual motion. In the basketball's case, it would be a parabolic curve leading from hands to hoop.

The least action principle is marvelous in that it recasts classical physics on an intuitive basis. Everything in the universe tries to find the optimal path from start to finish. In a kind of survival of the fittest, the most expedient trajectories win out. Like marks on a school report card reflecting good or bad behavior, the action quantifies the efficiency of each path, singling out the best for special notice. That turns out to be the route objects actually take in the classical realm.

LIBATION AND INSPIRATION

Despite having many insights into the application of quantum methods to the absorber theory, Feynman struggled to find the precise mathematical machinery for doing so. None of the existing techniques could relate things distantly separated and acting on each other remotely. Resounding in his tired brain was the need for a whole new approach. He knew that he had to start completely from scratch and somehow rebuild quantum physics using a least action principle—but how?

On posh Palmer Square, across Nassau Street from campus, was one of Princeton's most famous watering holes, the Nassau Tavern (now the Nassau Inn). Taking a break from his musings, Feynman decided to attend a beer party there. As luck would have it, he'd meet someone at the event who possessed the missing piece to his quantum puzzle.

Herbert Jehle, an exiled German physicist, introduced himself at the party and casually asked Feynman about his research. Jehle had just fled the notorious Gurs Internment Camp in southern France where the Nazis incarcerated him for his pacifist and antifascist views. Just arrived in the United States, he was briefly visiting Princeton.

After Feynman explained what he was doing, Jehle thought hard and remembered a key paper by Dirac, "The Lagrangian in Quantum

Mechanics," published in 1933. The article wasn't commonly known (at least among American theorists) at the time, as it had appeared in a relatively obscure journal, *Physikalische Zeitschrift der Sowjetunion*. In it Dirac had demonstrated how the transition between any two quantum states could be expressed as the product of special mathematical factors, called the "generalized transformation functions," which depend on the action—related to the Lagrangian—from point to point. Recall that Lagrangians are determined specifically from the difference between kinetic energy (energy of motion) and potential energy (energy of position) at each locale. These energies depend on the dynamical variables (position and momentum) in question. The generalized transformation functions convert the action into factors that, when multiplied, gradually convert an initial quantum state into a final quantum state via a chain of intermediate steps. Forming such a product corresponds to slicing up any quantum process into infinitesimal transformations, much liking dividing a movie reel into a series of still shots.

Although this method is rather technical, we might illustrate its key idea by means of analogy. Let's represent a quantum process by rows and rows of upright dominoes, standing in for the quantum states, spread throughout an uneven terrain, symbolizing the dynamical variables. Just as the dynamical variables set the generalized transformation functions that gradually convert quantum states from one to the next along a particular path, we might imagine the unevenness of the terrain inducing a cascade of falling tiles—toppling one after another in succession along a certain course. The chances of each tile tumbling in a certain way have to do with the specific landscape at that point. Similarly, dynamic variables of a quantum state at a given point in time set the odds of it "toppling" in a certain way to form the next state, and so forth—leading to a ripple of quantum transformations from start to finish.

AN ACTION-FILLED ADVENTURE

Feynman grabbed Jehle's suggestion and ran with it to his long-awaited goal. Perusing Dirac's paper, he immediately saw how its Lagrangian methods were ideal for quantizing the Wheeler-Feynman absorber theory. By formulating the theory in terms of an action principle and defining the classical trajectory as the least action path, he could frame it as a range of quantum possibilities. In our analogy, that is like demonstrating how the densest track of fallen tiles—corresponding to the likeliest classical path—is surrounded by a plethora of other, less-piled-up domino trails, representing less probable routes. In other words, while the dynamic landscape tends to steer the dominoes a certain way, laying out a favored path when they fall, chance allows for other, less likely possibilities. Similarly, while the classical trajectory is favored, Feynman developed a way of showing how it is but the peak of a tapered mound of quantum alternatives. He could thereby inject quantum physics' fundamental uncertainty into the absorber theory.

As Feynman realized, quantum haziness mandates that interactions cannot be confined to one particular avenue, which would be like trying to channel a thundercloud through a wire. Like thunderclouds, quantum positions are amorphous. Yet, sometimes lightning strikes, illuminating the most efficient way charges might travel. It is not the only way, just the likeliest. Similarly, within the "cloud" of a quantum process, one might identify an optimal route. That most efficient path—a lightning flash within a thundercloud—marks the classical trajectory.

To stir up such quantum murkiness, for any pair of interacting particles, Feynman determined every conceivable series of interactions that might connect them. His tally included not just the classical trajectory and other relatively likely paths but also those that seemed very indirect and improbable. The possibilities were limitless. At face value, all were equal. Yet, as in George Orwell's *Animal Farm*, some must be "more equal" than others.

To ensure that the classical trajectory turned out, in the end, to be the most probable, in his theoretical method Feynman weighted each path according to its likelihood, established by the generalized transformation functions that he found in Dirac's paper. For each trajectory, following Dirac's techniques, he set down dynamic variables at every point in time, calculated the corresponding Lagrangians, shaped these into transformation functions, and multiplied them together to represent an entire chain of events. Then, by summing these possibilities and applying the least action principle, he showed how the classical path is singled out as the likeliest of all. Feynman dubbed this special quantum addition a "path integral."

Feynman's method beautifully connected the principle of least action with Fermat's principle that light signals travel along straight lines in order to take the least amount of time. Therefore it naturally shows why classical signals between electrons follow the path of light cones. The transformation function for each path acts as a phase delay factor, indicating how much signals lag when they travel along that route. Because there is a multitude of paths, there is a range of delays. Just as in the interference of light waves, the various signals are added together to form an overall wave. In the weighted sum of paths, those signals following the most efficient route would have similar phases, constructively interfere, and thereby contribute the most. Following Fermat's principle, such an optimal trajectory would be along the linear path taken by light. In that manner, Feynman brilliantly showed how his quantum technique could reproduce the classical methods he developed with Wheeler within the limit of greatest likelihood, while allowing for a gamut of less probable paths constituting the haze of quantum alternatives. In other words, it blurs the narrow classical interaction into a broader quantum range of contributions.

Wheeler found Feynman's path integral method truly remarkable. It made the otherwise abstruse mechanisms of quantum dynamics seem as simple as basic optics. He thought that it connected classical and quantum theories in an even more natural way than either Heisenberg's or Schrödinger's formalisms. He counted his blessings for the good fortune of having such an extraordinarily original thinker as a student. To help promote what he saw as a revolutionary new concept, he decided to nickname it "sum over histories." As physicist Kenneth Ford, another of Wheeler's students and coauthor of his memoirs, recalled, "Wheeler told me that he—ever in search of striking names and phrases—gave that name to what Feynman had been calling the path-integral method."

As his excitement for Feynman's method grew, Wheeler thought he might even be able to persuade Einstein of its brilliance. He stopped by Einstein's house again and had a deep discussion with him in his upstairs study. Wheeler asked if Feynman's novel technique might persuade him to drop his opposition to quantum theory. However, Einstein, eyeing the theory's chance component, couldn't be swayed.

"I can't believe that God plays dice," said Einstein. "But maybe I've earned the right to make my mistakes."

THE SINGULAR LIFE OF AN ELECTRON

At some point while developing his path integral method, Feynman was resting up at the Graduate College when his dormitory phone started ringing. He picked up the receiver and heard an excited voice on the line. It was his crazy supervisor, with another of his wild ideas.

Wheeler relayed to Feynman that he had figured out why all the electrons ever detected have identical charges, masses, and other properties. There is only one electron in the universe, he explained. All the electrons we see are simply the same one racing forward and backward in time—ricocheting through perpetuity like a racquetball in a court. That is why every electron appears the same.

We think there are many electrons instead of one because we're witnessing only a single moment in time—a mere slice of reality. In that snapshot, we're seeing the same electron in all its incarnations, occupying many different places as it zigzags through eternity. These versions might interact with each other much as a thread, sewn again and again through a button, might be tied together.

We might envision such a situation via cinematic analogy by recalling how the protagonist Marty McFly in the film *Back to the Future II* needed to return to the setting and era he visited in *Back to the Future*: Hill Valley in 1955. Consequently, in that locale, there were two versions of him, constituting different loops in the thread of his timeline. Imagine if he did that again and again in countless sequels. Ultimately the town would house myriad versions of Marty McFly.

Visionary writer Robert Heinlein, in the story "All You Zombies," pondered a related situation in which a character became his own mother, father, and friend—by looping through time repeatedly, undergoing gender changes, and interacting with him- or herself. If traveling back in time were possible, such bizarre situations might be conceivable.

Wheeler imagined a sole electron in the starring (and only) role of its own time-travel saga. He hypothesized that at any given time and place, we might witness the many sequels of that time-traveling electron's adventures. Again and again it would cut through our slice of reality, until it would seem like the universe was full of such particles. Yet it would be flying solo.

Each time that singular electron traveled backward in time, Wheeler noted, its charge would seem to flip because our perceptions are future directed. Therefore, we would perceive it as moving forward in time. Mathematically, according to the Dirac equation, a negative charge heading into the past looks like a positive charge aimed for the future. Reversing both charge and temporal direction (along with spatial direction if the electron is moving through space) happens to result in the same solution. If we detect it at that stage, we call it a positron. Whether to consider it a backward-traveling, negatively charged electron or a forward-traveling, positively charged positron becomes simply a matter of semantics.

Feynman was dubious at first. If Wheeler was right, where were all the positrons? If electrons zig and zag through time, transforming into positrons and back again into negative particles, researchers should detect equal numbers of electrons and positrons. On the contrary, positrons were far rarer.

Having reflected on that imbalance, Wheeler offered an answer. Most of the positrons in the universe, he theorized, were embedded in protons. Anticipating, in some ways, the idea of quarks, he proposed that protons might be composite particles, hiding positrons in their innards.

Feynman relaxed into the idea of positrons as backward-traveling electrons. It offered a simple way of conceiving of positrons, fit the equations well, and made calculations easy. He didn't think too much at that time about the notion of protons as amalgams. Much later in his career, though, he would explore the subject of proton constituents—not positrons, but something he would call "partons."

Neither Feynman nor Wheeler spent too much time after that phone discussion pondering the concept that all electrons were the same particle. There was no obvious way to test that wild idea. More practical considerations soon leaped to the forefront.

DREAMS AND NIGHTMARES

From "everything is scattering" to "everything is electrons" and onward to "one electron is everything," Wheeler's active mind flitted like a butterfly from the nectar of one tantalizing idea to another. Taking a nurturing sip, it would drink its fill and move on to other delicious notions. It was often too restless to wait for experimental confirmation.

Feynman came to know his mentor's style very well and didn't mind it one bit. After all, Max Planck's concept of the quantum, Einstein's relativity, Bohr's wave/particle duality, and Heisenberg's uncertainty principle all seemed completely bizarre too—until they became mainstream. Feynman knew that Wheeler also had a cautious side and always kept within the bounds of physical law. He usually confined his most speculative ideas to discussions and notebooks until he found a way to defend them with meticulous calculations and elaborate reasoning. Both physicists agreed that scientific research was a serious business requiring careful justification. Yet sometimes, when progress in a certain area was stymied, one had to think big. Wheeler loved that "big picture" side of physics.

"I have never been too busy to dream," Wheeler once recounted. "Dreaming of what might be—of how the world is put together and how its parts interact—provides necessary sustenance for my brain, as nourishing as any calculation."

Wheeler's career would remain a teeter-totter of wild dreams and more practical considerations. As in the case of the watch incident, he always tried to make the best use of his time—finding the right balance between fanciful luxuries and somber necessities.

Grim news on December 7, 1941, certainly tipped the scales toward the latter. Japanese bombers launched a massive surprise attack on Pearl Harbor Naval Base in Hawaii, the first on American soil in generations. The next day, the United States declared war on Japan. A few days later Germany and Italy, allies of Japan, declared war on the United States, which immediately reciprocated with its own declaration of war. Catapulted suddenly into World War II, Americans braced themselves for fierce fighting in Europe and the Pacific.

Wheeler recalled Bohr's startling announcement, almost three years earlier, of the German fission discovery. His European-born colleagues and friends, such as Einstein, Wigner, Fermi, Leo Szilard, and Edward Teller had been far more concerned than he was. Wheeler had believed at the time that any warfare would be confined to nations overseas. Historically, Europe seemed perpetually embroiled in various political struggles and conflicts.

American involvement, however, changed everything. The United States would need to win the war as rapidly as possible. That meant developing superior weapons. Clearly, the Allies had to take every step to prevent the Axis powers from harnessing the energy of the atomic nucleus.

Wheeler would soon discover that President Franklin Roosevelt had already decided to do just that. On December 6, the day before Pearl Harbor, the US Office of Science Research and Development had begun a project to investigate atomic power, headed by a group called the S-1 Uranium Committee, soon jointly managed with the War Department. Later called the Manhattan Project, it would grow throughout 1942 and 1943 into a mammoth enterprise with the goal of investigating atomic weaponry, producing fissionable material, and constructing bombs if feasible. Wheeler and Feynman would both play major roles.

Little more than two years after Wheeler and Feynman started working together, it had become clear that their time discussing theoretical physics at Princeton would be very limited. Called to contribute their expertise to the war effort, they would need to work on military projects in various parts of the country. In an alternative reality, perhaps they would have had the leisure to explore "crazy ideas" further in the cozy settings of Fine Hall and Palmer Lab. However, fate would whisk them away to places very far from paradise.

CHAPTER THREE

ALL THE ROADS NOT TO PARADISE

Time forks perpetually toward innumerable futures.

—Jorge Luis Borges, "The Garden of Forking Paths"

Some of the most riveting conversations about any aspect of history start with the simplest of questions: What if? What if the Norman invasion of England had never occurred? What if the South had won the American Civil War? What if Leon Trotsky had led the Soviet Union in the 1930s instead of Joseph Stalin? A room filled with imaginative guests supplied with ample wine and cheese might debate the implications of countless scenarios for hours.

Of course, no one really knows what would have happened in such alternative histories. Therefore, given that no one can prove or disprove any hypothesis, such debates rarely have winners and losers. They are intellectual exercises, pure and simple.

In the modern era, many such alternative history discussions center on World War II. The numerous critical decisions made on both sides offer plenty of opportunities for second-guessing. For example, Adolf Hitler voided a pact with Stalin and invaded the Soviet Union. In doing so, he awoke a mighty sleeping giant. If he hadn't committed such a betrayal, perhaps he could have held onto some of his other gains.

On the Allied side, arguably the most controversial decision has been the dropping of atomic bombs on the Japanese cities of Hiroshima and Nagasaki. That nuclear horror took hundreds of thousands of lives. Some critics argue that the decimation of such large, defenseless civilian populations was cruel and unnecessary. Perhaps military targets could have been chosen instead—or a warning or demonstration conducted. Others rebut that the actions prevented a bloody continuation of the war in the Pacific theater, which might have involved a high-fatality ground troop invasion of Japan. The number of lives lost in those two cities, these pundits argue, pales in comparison to how many would have fallen in a prolonged conflict.

When the Manhattan Project began, it clearly strove to counterbalance possible nuclear efforts in Nazi Germany. No one knew whether it would succeed, let alone result in such massive casualties. As it turned out, Nazi Germany made almost no progress toward a nuclear bomb throughout the war. Had the United States known that in advance, perhaps it would have channeled the considerable funding and manpower for the project elsewhere. What if?

On the human level, the choices made at any critical juncture set the course of the future. No one can know what would have happened had circumstances been otherwise. But imagine if an alien race somehow possessed the ability to envision all possible scenarios. As if flipping through channels on a cosmic television, the hypothetical beings could view worlds with Hitler and without Hitler, with Franklin Roosevelt and without him. Suppose the alternatives are really out there but inaccessible to us. Our "subscription" just allows us one channel among countless possibilities. Would the idea that everything that might happen *does* happen in some parallel strand of time rob history of its seeming inevitability?

Richard Feynman's sum-over-histories proposal introduced such a forking view of time at the quantum level. Any interaction between particles happens in all conceivable ways, not just one, as long as the laws of physics are obeyed. To compute the total outcome, one's "quantum television" must be able to pick up every possible channel. Only by tracking and including all viable paths might one achieve a complete picture of reality.

THE RHYTHMS OF LIFE

Historically, there have been many different models of time. For the ancient world, the most prevalent was the idea of cycles. Perhaps that is not surprising. The circadian rhythms of our bodies, the repeated patterns of astronomical bodies, and the ceaseless progression of the seasons each suggest that time is cyclical. The continued popularity of astrology and the concept of reincarnation serve as testimony to the persistence of such a notion.

Nature's beat manifests itself in countless ways. The cosmos offers cycles within cycles: from the daily rotation of the Earth, to its steady annual movement around the sun, to the combined motion of the sun's orbital family around our galaxy's central hub. Day follows night; thaws follow frosts. The moon's powerful pull sets tidal patterns. Each celestial body marches to a rhythm set by its own balance of energy.

Living creatures respond to these cyclic patterns with their own periodic behavior. Birds migrate, bears hibernate, and salmon swim upstream each year to spawn. Human life similarly marches in step. We awaken, eat, and sleep at regular intervals—even if we live or work in places untouched by sunlight. Try as we might to ignore such circadian rhythms, our bodies betray us with wakefulness, hunger pangs, or fatigue.

Given the pull of daily and seasonal patterns, perhaps it is not surprising that most ancient cultures believed time to be fundamentally cyclic. From Mayan calendar round stones to the Chinese yin-yang icon, from the Egyptian ouroboros (the image of a serpent devouring its own tail) to the Indian wheel of Asoka, we find references to cycles in symbolism around the world. Most venerable civilizations followed calendars that included not just days, months, and years but also much longer periodic intervals of cosmic destruction and rebirth.

For example, according to the Puranas, Hindu sacred texts written in Sanskrit starting in the fourth century AD, the world is built on cycles within cycles of varying durations. History proceeds through a repeating series of epochs, called *yugas*, each many thousands of years long. In some accounts these are embedded in longer intervals, called *mahayugas*, which persist for millions of years. Each of these in turn forms part of an even greater era, called a *kalpa*, which lasts for billions of years. Celestial events, such as the alignment of planets, and terrestrial calamities, such as floods and fires, mark the transitions between various cycles.

Cyclic time is eminently repeatable, reversible, and deterministic. As poetically expressed in Ecclesiastes, "To every thing there is a season." Wait long enough, and each phase of a cycle comes around again.

Yet cyclical time cannot be the full picture. Numerous features of the natural world suggest one-way arrows of linear time. One such arrow arises in thermodynamics: the study of heat and energy. According to the second law of thermodynamics, developed by German physicist Rudolf Clausius in the mid-nineteenth century, for any closed process a quantity called "entropy" must either stay the same or increase but never decrease. Entropy reflects how much energy in a system is unavailable for work; the higher the entropy, the greater the amount of wasted energy.

Entropy is also a measure of a system's lack of uniqueness: the more exceptional the system, the lower its entropy, and vice versa. Typically, ordered systems have lower entropy than those that are disordered. "Ordered" in this context represents the uniqueness of the arrangement of particles in the system.

For example, it is much harder to assemble a snowflake with a particular pattern than a puddle from a set of water molecules. The snowflake's intricate arrangement of constituents is much less common than the puddle's sloshy liquid state. Therefore the former has lower entropy than the latter. Following the second law, a snowflake on the ground might readily melt into an amorphous fluid, but ground-level water never spontaneously transmutes into varied snowflakes.

Lord Kelvin (William Thomson) postulated that the entropy of the universe will gradually increase over time, leading to less and less usable energy, until the cosmos reaches a completely inert state called "heat death." By that time the furnaces (which we know today to be nuclear) that heat the cores of stars will cease to function. Their outer envelopes will evaporate or explode, leaving behind inert interiors. Even those cold remnants (which we now know to be white dwarfs, neutron stars, and black holes, depending on the stellar mass) will gradually cede their energy to space over time, until the entire universe is completely frigid and lifeless. The signpost pointing toward that bleak future is called the "thermodynamic arrow of time."

Contrasting with this arrow is the direction of evolution. Biology tells us that, at least on our tiny planet, processes occur that result in enhanced, not diminished, complexity. Life's development over billions of years shows how, through the mechanisms of variation and natural selection, simple organisms evolve into highly complicated beings, such as dolphins, chimpanzees, dogs, and their bipedal caretakers. Even the cynical would agree that humans are much more complex than amoebas. Humans have the remarkable ability to understand themselves, anticipate future possibilities, reshape the environment, and map out the cosmos, among other advanced attributes. Technology, the product of human ingenuity, similarly becomes more advanced from era to era. Therefore evolution offers a progressive arrow of time.

Which arrow of time will win in the end, the thermodynamic or the evolutionary? Unless things change, it will be the former. Life requires a continuous influx of ordered energy—from the sun or other sources of usable fuel. Eventually all living things lose their capacity to maintain their equilibrium and die. Given the fragility of life on Earth, the universe will probably far outlast it. However in some speculative scenarios—notably in Isaac Asimov's science fiction masterpiece "The Last Question"—advanced civilizations eventually learn how to reverse the second law, resulting in the triumph of life's organizing powers over the tendency toward decay.

The expansion of the universe offers yet another arrow: cosmological time. Dating back to discoveries by Edwin Hubble and others that the bulk of galaxies are moving away from ours, there is considerable evidence that space has grown enormously over time. No one could mistake the universe today, with its myriad stars and galaxies spread out over billions of light years, for its compact primordial precursor. A chart of spatial expansion offers a clear cosmological arrow.

All of us, even those unaware of scientific arrows of time, have a strong sense that time moves forward. Our conscious awareness seems to propel us ever onward, from birth until death. In our familiar experiences, causes always seem to precede effects.

Although time is intangible, we cannot seem to escape its incessant stream. What causes time's apparently ceaseless flow? Is our sense of forward passage an illusion, like the perception of animation in a flipped series of still prints? Or is it a genuine phenomenon related to one or more of the other arrows? Regardless of its origin, whether illusory or real, consciousness represents yet another arrow of linear time.

Linear time has turned out, however, to be insufficient for describing nature at its fundamental level. As John Wheeler and Richard Feynman's work together demonstrated, certain processes in the quantum world seem to defy causality. Maxwell's equations yield signals that travel backward as well as forward in time; the Wheeler-Feynman absorber theory mixes them together. Wheeler's suggestion, which Feynman eagerly adopted, that positrons are electrons moving in an opposite temporal direction further decoupled particle physics from any semblance of an arrow of time.

TIME AS A LABYRINTH

Instead of a cycle or arrow, the sum-over-histories approach suggested a third distinct way of viewing time: as a labyrinth of ever bifurcating possibilities. Each point in time has many branches reaching into the future and roots digging into the past. These extensions twist, turn, and merge with others, only to split once more. In the quantum world, following only one of these limbs is insufficient. One must grapple with the tree, knotty as it is, in its entirety.

The term "labyrinth" dates back to one of the most compelling of all Greek myths, the story of the Minotaur, half man and half bull, trapped in an inescapable maze. In that tale, King Minos had assigned Daedalus, an extremely clever scientist, architect, and inventor born in Athens but living in exile in Crete, to design a home for the abominable creature. Daedalus built a vast complex of winding corridors, spiral staircases, lofty towers, and featureless rooms, which he called the Labyrinth, named for the sacred Cretan double-headed axe symbol. Upon completing this stony network, he placed the Minotaur in the center, assuring the king that the monster had no hope of escape.

On a trip to Athens, a dangerous bull had killed Minos's son. Seeking revenge, Minos decided to punish the Athenians by demanding that every nine years seven lads and seven maidens be sent to Crete as tribute. Once these youths reached Crete, they were placed in the Labyrinth, where, unable to escape, they awaited a horrible fate at the hands of the Minotaur.

The Labyrinth was so craftily designed that they couldn't possibly flee. Instead they wandered lost through its twisting corridors, complex passages, and endless rooms, not knowing where or when the monster would attack. After they had endured a prolonged period of despair, the creature would devour them.

Sympathetic to the fate of his fellow Athenians, a powerful young lad named Theseus decided to solve the maze and slay the Minotaur. Minos's young daughter, Ariadne, helped him in his endeavor. Advised by Daedalus, she gave him a thread to tie to the front entrance of the maze. Theseus entered, clutching the other end of the string in his hand, and searched for the Minotaur. Finding the monster asleep, he pinned it to the ground and beat it with his bare hands until it collapsed on the floor. He then painstakingly followed the guiding string back to the entrance, escaped, and proclaimed victory for the Athenians. Returning to Athens a hero, he later became its king.

Many scholars, such as the late semiotician Umberto Eco, have viewed the Labyrinth tale as a metaphor for humanity's attempt to map out and reproduce the complexity of the universe. Daedalus, in that interpretation, is the prototypical scientist. Eco pointed out in *Postscript to the Name of the Rose* that labyrinths might have varying degrees of complexity. While unicursal labyrinths have only one possible route, multicursal versions have many. The most complex type, rhizomic, would have an infinite range of possible avenues.

In a quantum labyrinth, Theseus would take not just one path to the center but a multitude simultaneously. Laying down an intricate web of threads, rather than a single one, he would explore various alternative routes. Along some of these he might arrive with enough vigor to kill the abomination inside. Along others, far twistier, he might be too tired to carry out the task. When the Athenians later recounted his story, they would need to address each of these possibilities, with somewhat more emphasis on the likelier choices that he might have made. Perhaps most days they'd tell the tale in such a way that Theseus made the wisest, most direct choices, but on others they'd picture him making foolish, less probable decisions that sent him to his doom. Instead of one classic history, there would be a quantum sum over histories, confusing students of Greek mythology forever more.

"THE GARDEN OF FORKING PATHS"

By sheer coincidence, in 1941, as Feynman was developing sum over histories, Argentine writer Jorge Luis Borges published "The Garden of Forking Paths," a remarkable short story with the theme of time as a labyrinth. Set during World War I, the tale is a murder mystery involving a Chinese spy named Yu Tsun who befriends and then assassinates noted British sinologist Stephen Albert. The sudden and unexpected turnabout reveals the capriciousness of time's stream. With time having so many possible branches, the story suggests, a chance event might deflect someone's life journey from an auspicious flow to a current of doom.

Yu Tsun ostensibly visits to consult with Albert about his ancestor, a learned Chinese governor named Ts'ui Pen. An expert on astronomy, mysticism, and mathematics, Ts'ui Pen had mysteriously resigned with the intention of writing a novel and building a labyrinth. Strangely, while a book was indeed written, no physical maze was ever constructed.

As it turns out, the labyrinth is the book itself, a chronicle in which, for any critical juncture, numerous contradictory outcomes occur. In one chapter a character is dead; in the next he is mysteriously alive. Two accounts of the same battle describe how demoralized troops sacrifice themselves for victory and how an invigorated army handily triumphs. The last page of the novel is the same as the first, suggesting continual rereadings and reinterpretations. Truly it is far more complex than any hedge maze.

After showing Yu Tsun the book, Albert explains its motivation:

> The Garden of Forking Paths is an incomplete, but not false, image of the universe as Ts'ui Pen conceived it. In contrast to Newton and Schopenhauer, your ancestor did not believe in a uniform, absolute time. He believed in an infinite series of times, in a growing, dizzying net of divergent, convergent and parallel times. This network of times, which approached one another, forked, broke off, or were unaware of one another for centuries, embraces all possibilities of time. We do not exist in the majority of these times; in some you exist, and not I; in others I, not you; in others, both of us.

Proving, in a way, his ancestor's thesis, that the universe is a network of random possibilities, Yu Tsun suddenly murders Albert. At first glance, the killing appears completely senseless. However it turns out that Yu Tsun is spying for Germany and wants to alert Berlin military command to the name of a city, Albert, which will be the next bombing target. He believes that the only way to get the name in the news is to assassinate someone with that appellation. The decision reveals the haphazard, labyrinthine nature of time. In another branch, Yu Tsun might be a lifelong friend of Albert, but in this one, he is compelled to kill him.

Today we would call a novel with numerous possible outcomes an example of hypertext. Thanks to the web, most of us encounter hypertext every day. When we read a news story and decide to click on a link that takes us to another site, we embark upon a journey through a textual maze of possibilities. We might wonder, after a time, why when starting with the intent to research the repercussions of World War II, we have ended up reading an article about Tuvan throat singers or bongo drumming. The choices we make guide our reading experiences along unique, personal avenues. Through our daily rambles on the web, where each array of links is a bifurcation of alternatives, labyrinthine time has become a familiar part of our lives.

RIFTS AND FISSIONS

One of the horrors of warfare, as depicted in Borges's story, is that it can turn friends into enemies. World War II divided the international physics community into opponents and defenders of the Axis powers—the former was a much more substantial group, but the latter included key figures, such as Werner Heisenberg. While staunchly opposed to the Nazis, he decided to remain in Germany and work on scientific endeavors. He was placed in charge of a group investigating the prospects for nuclear energy and weapons. The team received few resources and, as mentioned, would make little progress.

Heisenberg's actions greatly alarmed Niels Bohr and his other former colleagues. Despite their prior respect for Heisenberg, many came to despise him. Others clung to the belief (which he tried to promote after the war) that he had deliberately dragged his feet in the German bomb effort.

The Manhattan Project, in contrast, constituted an unprecedented gathering of resources, know-how, and personnel that would encompass numerous sites throughout the United States. At first, its hubs were research laboratories at the University of Chicago and Columbia University in New York (the latter site gave the project its code name). Ernest Lawrence's Radiation Laboratory at Berkeley, where cyclotrons were developed, and a lab of the same name at MIT would play major roles. Later, major secret centers, housing thousands of researchers and staff, would be established at Los Alamos in New Mexico, Oak Ridge in Tennessee, Hanford in Washington, and elsewhere.

In the time between Bohr and Wheeler's key paper on fission and the US entry into the war, nuclear chemist Glenn Seaborg had used a cyclotron to produce the first samples of the artificial radioactive element plutonium. As calculated by Bohr and Wheeler, plutonium-239 and uranium-235 each showed promise as possible ingredients for a fission bomb. After the paper was published, Wheeler had continued to research nuclear structure, along with his older Princeton colleague Rudolf Ladenburg and Ladenburg's PhD student Henry Barschall. However, his major focus had been on scattering and particularly on his collaboration with Feynman. The American entry into the war changed all that. His nuclear skills were very much in demand.

In January 1942, Wheeler was invited to the University of Chicago to investigate the idea of building a production reactor to create sufficient supplies of plutonium. Under Arthur Compton's direction, Chicago had rapidly become the plutonium research center for the war effort. Enrico Fermi had been recruited there from Columbia, where he had already achieved a chain reaction in a nuclear pile. Eugene Wigner also spent time there. Code-named the Metallurgical Project, the Chicago group needed more bright minds to determine the most efficient way to generate plutonium in uranium reactors and isolate substantial quantities for use in a bomb.

Soon, the demands of the Chicago project forced Wheeler to abandon his work with Feynman altogether. He profoundly regretted having to set his theoretical dreams aside, for a time, to focus on military research. Nevertheless, he felt duty-bound to do all he could for the war effort.

Meanwhile, Robert R. "Bob" Wilson, a young Princeton researcher and former student of Lawrence, approached Feynman with a hushed request. A secret military project then going on at Princeton had to do with separating uranium isotopes for bomb development.

Wilson asked Feynman if he'd be interested in learning more. If so, he'd be welcome to attend a meeting at 3:00 p.m. that day.

Feynman, by that stage, was getting ready to finish his thesis project and earn his doctorate. He had passed the oral qualifying exam, which included a question about the order of colors in rainbows that, because he didn't remember the answer, he had to talk his way through carefully. Also, he had written up a twenty-seven-page explanation of the action-at-a-distance theory, in its classical and quantum forms, which was possibly ready for publication and could form the kernel of a thesis. He was looking forward to the moment when, PhD in hand, he could finally marry Arline and get a job—ideally an academic position where he could teach, but alternatively at a research company such as Bell Labs.

Consequently, Feynman's first reaction to Wilson's request was to say no. Completing his thesis was an absolute necessity, he explained. He needed to graduate first; then he could think about other things. Besides, he thought, the last thing in the world he wanted to do was work on weapons. That's not why he went into physics.

He returned to his room and started to work on his manuscript. Then he thought carefully about the devastating repercussions of an Axis victory. What if the Nazis built a nuclear bomb and used it to attack the Allies? How could he live with himself if they destroyed cities with colossal atomic explosions? With that horrific image in mind, he tucked his thesis into a desk drawer.

The meeting was brief but informative. Wilson and other researchers spoke about their goal of separating uranium isotopes using an electromagnetic device called an isotron. After the meeting, in a room in the building with a rolltop desk and plenty of paper, Feynman immediately became immersed in the calculations. He decided to make the project a priority.

Soon, thanks in part to Feynman's computations, Wilson's group managed to produce substantial quantities of uranium-235. They sent these samples to Columbia and elsewhere for testing. A steady stream of scientific visitors stopped by Princeton to examine its separation facility. Whenever a new visitor arrived, Wilson called on Feynman to explain the technical details of the separation process.

THE RUSH TO A DEGREE

The spring semester of 1942 was an extraordinarily busy time for both Wheeler and Feynman, with both involved in separation projects for different fissionable materials. The key difference was that Wheeler already had a stable academic position, whereas Feynman did not. The physics department chair, Henry "Harry" Smyth, a specialist in nuclear physics, was a member of the national S-1 Uranium Committee and thus very supportive of Wheeler's nuclear efforts. Smyth granted him a great deal of flexibility in juggling his teaching duties and other responsibilities so that he could travel back and forth between Chicago and Princeton.

Concerned that Feynman's nuclear work would distract him from completing his PhD, on March 26, Wheeler wrote from Chicago urging him to finish his thesis: "Both Wigner and Ladenburg feel with me that you have done more than enough for a thesis. . . . I would advise you very strongly to write up what you have in these remaining few weeks before you get into the situation in which I now find myself, where there is absolutely no time to work on our theory of action at a distance." Wheeler finished the letter with a personal note: "I hope Arlene [*sic*] is feeling better."

Unfortunately Arline wasn't doing well at all. The burden of her situation weighed heavily on Feynman. Her condition had grown more severe, and she regularly coughed up blood. She was at such an advanced stage of tuberculosis (TB) that doctors could do little beyond isolating her in a sanatorium.

Richard still wanted to marry his beloved Putzie—at the very least to help take care of her and offer emotional support until she got better. His parents were very concerned, however, that he'd be making a huge mistake. His dad, for instance, had high hopes for his career and worried that marriage to a woman with a chronic, infectious illness would harm his chances of getting a job. His mother was a nervous wreck, thinking about everything that could go wrong.

On his father's advice, he asked Smyth what to do. The department chair didn't think he'd have a problem getting a position. Just to be certain, Smyth referred him to Dr. Wilbur York, the university physician.

Dr. York was very helpful, laying out the challenges of marriage to a woman with infectious tuberculosis. His calm, rational demeanor put Feynman at ease. Marriage, he noted, would help Arline feel more emotionally secure, aiding in her recovery. The doctor reassured Feynman that, given the many precautions taken at a sanatorium, he had little chance of catching her illness. He could certainly safely hold an academic job and teach students.

On the other hand, Dr. York emphasized, he would have to make absolutely sure that she didn't get pregnant. Feynman assured the doctor that there was no chance of that happening. Finally, the doctor took him aside and sternly advised that TB wasn't always curable. Feynman replied that he knew the risks but wanted to marry Arline anyway.

In light of Wheeler's admonition to finish his thesis, Feynman redoubled his efforts. While he figured that Wheeler would probably pass him on what he had done already—the classical action-at-a-distance results and some of the quantum stuff—he felt obliged to complete the latter so that it was a coherent body of work. Otherwise, if he ever wanted to return to the topic, his scattered notes might seem like nonsense, and he might never be able to reproduce what he had achieved.

To conclude his thesis work, Feynman asked for and received about a month's leave from the uranium-separation project. Exhausted from the nuclear computations, he spent the first few days of the mini-sabbatical just lying on the grass and staring at the sky. That break was all he needed to complete his quantum calculations and write up the rest of his thesis. Wheeler was back at Princeton, so it was perfect timing. Feynman handed copies of the finished thesis to him and Wigner, who was also on his committee. They enthusiastically approved it and granted him the PhD.

I THEE WED, IN SICKNESS AND . . .

On June 16, Feynman donned a traditional cap and gown and marched in Princeton's commencement ceremony. His parents were beaming. They were used to his piling up accolade after accolade, including medals at math contests, but a PhD from Princeton in only three years was major cause for celebration.

Because of the war work, he knew that he'd need to remain in Princeton during most of the summer and perhaps beyond that. He also had an alluring job offer: a visiting associate professorship at the University of Wisconsin. It would start in September if he somehow finished the war work (or the war miraculously ended). Otherwise, he could defer it for a year. He accepted the position based on those flexible conditions. Indeed he ended up having to defer—and ultimately never took the job.

On top of his other responsibilities he had one more sworn duty—to a certain Miss Arline Greenbaum. After many years of engagement, it was time to tie the knot. While Arline's family supported the couple's decision, Feynman's parents remained very nervous because of her ill health.

Thinking her son was marrying simply out of obligation, Feynman's mom urged him to reconsider. Recognizing his seriousness, she became frantic to stop him from making the mistake of his life. Getting married is not like eating spinach, she wrote to him. It shouldn't be a burden. He'd be taking on untold medical and emotional costs, which could potentially divert him from achieving his career goals. She urged him to stay engaged until Arline was completely better.

Feynman wrote back that he had already made up his mind. He explained that he loved Arline, wanted to take care of her, and saw no conflict between marrying her and pursuing an active career in physics. In fact, her love and support might help him focus on his research, he reasoned. In short, he reassured his mother that he'd be just fine, whatever happened.

Richard and Arline were wed on June 29 in a New York courthouse. It was a simple ceremony, with no family members present. Their honeymoon was a simple Staten Island ferry ride, a one-nickel-fare crossing that many commuters took each day. Afterward, he took his new wife right to Deborah Hospital in Browns Mills, New Jersey, a well-known sanatorium conveniently located some thirty-five miles southeast of Princeton, where she would be staying. Nestled in the sylvan Pine Barrens, it had a relaxing setting, fragrant with fresh pine air. The proximity would enable Richard to visit as often as possible until Arline was well enough for the married couple to live together.

Ironically, a better treatment for Arline's ailment was geographically close at hand. At Rutgers University Agricultural School, about fifty miles from where she was situated, a compost heap was located

Richard Feynman and Arline Greenbaum Feynman on vacation in Atlantic City, New Jersey (*Source:* AIP Emilio Segre Visual Archives, Physics Today Collection, gift of Gweneth Feynman).

on a campus farm. No ordinary pile of muck, it contained the secret for curing TB.

There, on August 23, 1943, about a year after Arline checked into the sanatorium, graduate student Albert Schatz, working under the supervision of Dr. Selman Waksman, would analyze a soil sample and discover the antibiotic streptomycin, which would turn out to be extremely effective against TB. By the late 1940s, it would be on the market as a wonder drug that cleared TB symptoms in months and eliminated the need for sanatoriums. Deborah Hospital, for example, would switch its mission to other heart and lung conditions.

In a Hollywood ending or benevolent parallel reality, the Feynmans would happen to meet Schatz somewhere in New Jersey. Arline would try the miracle cure and stop coughing; her lungs would heal, and the newlyweds would live happily ever after. Alas, it was not to be. By the time of Schatz's discovery, Arline and Richard would be in New Mexico, where Richard would work at the secret nuclear facilities in Los Alamos. Sometimes the harshest two words are "what if?"

SECRETS AND CONFIDENCES

By late 1942, under the scientific directorship of J. Robert Oppenheimer, universally known as "Oppy" (also spelled "Oppie"), and the military leadership of General Leslie Groves, the Manhattan Project had grown into an enterprise too large and active to keep under wraps at major universities such as Berkeley, MIT, Princeton, and Chicago. To consolidate its research in an isolated location, the US government decided to purchase the Los Alamos Ranch School and convert it into a top-secret nuclear laboratory. By early 1943, the government occupied the site and began to repurpose it for military research use.

When Feynman learned that his research group would be moving from Princeton to Los Alamos, one of his foremost concerns was Arline. In March he wrote to Oppy and his Berkeley associate J. H. Stevenson inquiring about a hospital or sanatorium relatively close to the secret site (code-named "Project Y"). Oppy and Stevenson offered Feynman several options. He chose Albuquerque's Presbyterian Hospital, noted for TB treatment.

Arline remained remarkably upbeat and enthusiastic, despite her condition. She looked forward to the long train ride together, from Princeton to Chicago, and then onward to Santa Fe, New Mexico. Dreaming about soon setting up a real household with Richard—at least spending more time together—she hoped the move augured a better life for them. With its dry, desert climate, New Mexico was well known as a place where many TB patients went to recover. With all her heart, she wanted to get better and become the caring wife her Richard deserved. Perhaps they might even have a child, she thought, when the time was right.

To allay public suspicion, most members of the Princeton group left for New Mexico from various other train stations. Their luggage, however, was sent from Princeton. The Feynmans decided to be different and depart from the Princeton station itself. Because they seemed to be the only couple traveling to New Mexico, they were amused by the idea that the train officials would think that all the Princeton baggage belonged to them.

Enjoying a cross-country journey in a relatively luxurious carriage, the Feynmans finally had something like a real honeymoon. Arline had no idea what Richard would be doing, only that the project was secret. All the mail sent to Los Alamos was addressed to Post Office Box 1663, Santa Fe, New Mexico, where it would be collected, censored, and distributed to the researchers and other workers.

Feynman was assigned to the lab's Theoretical Division, headed by Cornell professor and nuclear expert Hans Bethe. Bethe had written the "bible" of nuclear theory: a three-part article delineating the essence of what was known about the field. Coincidentally, around the time Feynman arrived, many of the other theorists were away. Seeking a sounding board, Bethe stopped by his office several times and was highly impressed by his quick, seat-of-the-pants insight into what would and wouldn't work. "I realized very quickly that he was something phenomenal," recalled Bethe. "I thought Feynman perhaps the most ingenious man in the whole division, so we worked a great deal together."

Bethe and Feynman's loud but friendly disagreements over solutions to various technical issues related to nuclear bomb design became legendary at the lab. Bethe was very systematic and laid out his arguments point by point, with ample mathematical evidence, like an adept attorney making a case. Whenever Feynman thought Bethe was wrong, he'd suddenly blurt out something like, "No, no, you're crazy!" or "That's nuts!" Bethe would patiently continue, only to be interrupted vociferously again and again. Other researchers in the vicinity couldn't help but overhear Feynman's rebukes, which they generally found hilarious—especially given Bethe's esteemed status.

Bethe and Feynman both had the habit while calculating of fiddling with tangled bits of copper or plastic, which they called "thinking toys." One day Feynman, as a joke, exchanged his own toy with Bethe's. The result was remarkable. Feynman found that he grew more measured and systematic, his thoughts and actions slowing down. Bethe, on the other hand, appeared more animated and started to make wild gestures. Their colleagues laughed when they saw that Feynman and Bethe's much parodied personalities had briefly flipped.

While continuing to work closely with Bethe, Feynman also came to know Oppy very well. Oppenheimer similarly thought Feynman phenomenal. By the end of 1943, both Bethe and Oppy were already vying behind the scenes to snag him for their respective universities after the war. Each saw his potential as a gifted professor with a promising career.

Recommending Feynman's appointment to Berkeley's physics department, Oppenheimer wrote to Raymond Birge, its chairman:

> He is by all odds the most brilliant physicist here, and everyone knows this. He is a man of thoroughly engaging character and personality, extremely clear, extremely normal in all respects, and an excellent teacher with a warm feeling for physics in all its aspects. . . .
>
> Bethe has said that he would rather lose any two other men from his department than Feynman from his present job, and Wigner said, "He is a second Dirac, only this time human."

As it turned out, Birge would hedge in making a preliminary appointment. Berkeley was not ready to make an offer. Bethe's department at Cornell acted much more quickly and promised Feynman a postwar position—which he accepted.

What specifically did Feynman do during his two years at Los Alamos? A better question is, what didn't he do? His fingerprints were on virtually every piece of equipment, his brilliance behind most of the important calculations. Whenever a new computer or other device arrived, typically disassembled, Feynman was often the first to rip open the box, take out the pieces, and put it together. He'd master its mechanisms inside and out—from wiring it up to interpreting its readings—and serve as the point person if anything went wrong.

One of Feynman's key early contributions concerned the behavior of fast neutrons emitted by uranium-235. They diffused (spread out) in a manner impossible to calculate precisely for realistic cases. Taking up the challenge, he devised a step-by-step procedure—in some ways akin to the summation methods he developed at Princeton—which he programmed into the primitive IBM computers. Programming in those days was a scrappy process involving tubes and wiring—just the kind of hands-on stuff he loved.

Later, during the assembly of the plutonium bomb, Feynman flexed his computational prowess again and again. He cranked out the details of what would happen at every stage of the bomb's workings, from the mechanisms for its detonation to its explosive capacity. His expert knowledge of math, computers, and particle physics contributed to the creation of the most powerful weapon the world had ever seen.

Though immersed in developing the machinery for mass death, Feynman had little time to think about the moral implications of creating such powerful devices. Rather, he focused on the technical challenges and the duty to help defeat Germany. Only later would he reflect on the profound ethical issues.

PARTNERS IN PRANKS

Feynman also strove to be a good husband to his ailing wife—to the best of his ability, given the distance between them and the limitations imposed by his job and her health. Serving as her confidante and her clown, he did everything he could to cheer her up. She reciprocated with overflowing love and ample silliness.

Thousands of miles away from their friends and family back in New York, the young couple let their goofy, mischievous sides run wild. When Richard visited Arline in the Albuquerque hospital, usually on weekends, she'd make up games and pranks to amuse him. For instance, they pretended that a stuffed elephant named "Snuggle" in her room was real. Taking the focus off her own situation, she'd report on how Snuggle was doing. They'd dream about having a real family life together someday, for which taking care of Snuggle was practice.

Often when heading over to Albuquerque, Feynman borrowed the car of close friend and fellow researcher Klaus Fuchs. Little did he know that Fuchs led a secret life. Five years after the war, Fuchs would be found guilty of spying for the Soviet Union. He would serve time in prison before emigrating to East Germany.

For Richard's twenty-seventh birthday, Arline had a special newspaper prepared with the mock headline "Entire Nation Celebrates Birthday of R. P. Feynman." She arranged for the delivery and distribution of copies to his group at Los Alamos. He was amused to see the fake papers posted everywhere and generating much laughter among his colleagues.

Seeing his sweet, clever Putzie as a true confidante and "partner in crime," he loved telling her about his zany antics. Without revealing any classified information, he bragged about what he could get away with in the supposedly high-security lab. For example, he'd sneak through a hole in the perimeter fence to ensure that there was no record of his absence in the logbook and chuckle as he pointed out the security breach. Sadly, her health continued to deteriorate, and she needed all the cheering up he could muster.

Feynman was particularly proud of his lock-picking abilities, which he had learned from a fellow graduate student at Princeton. Equipped with a paper clip, he could open virtually anything that used a key. Cracking open padlock after padlock, he'd bring the contents of filing cabinets to meetings and complain about how nothing was secure. When Edward Teller insisted that his own desk drawer, containing secret papers, was impenetrable, Feynman took up the challenge and removed its contents through a gap in the back of the desk. "Feynman seemed to be composed in equal parts of physicist and humorist," recalled Teller.

After the keyed locks were replaced with combination locks, Feynman felt compelled to protect his reputation. Following many months of fiddling with dials, listening to click patterns, taking notes on each safe, and reading books on the subject, he mastered the art of combination safecracking. He relished the look on his coworkers' faces when he could open tougher safes too. Like a consummate magician, he loved amazing people with his feats.

Nothing delighted Feynman more than for others to think of him as a completely normal guy—who happened to have incredible tricks up his sleeve, such as cracking safes, solving difficult puzzles, or performing ridiculously hard calculations. To his colleagues he'd be plain old "Dick," completely without pretensions. Aspiring to be one of the boys, he'd drink, smoke, and use salty language. One night, though, he got so drunk that he wrote to Arline promising that he'd clean up his act and stop drinking and smoking.

Feynman was clearly proud of his more creative pursuits, such as his prowess on the bongo drums, a passion that began at Los Alamos and would last his whole life. Teller recalled that he "played them for hours each night." "I can hardly think of that earliest period without recalling the sound of bongo drums," said Teller.

A COMPTON-SCATTERED FAMILY

Wheeler's life during the Manhattan Project was very different from Feynman's. For one thing, he had three young kids. Being away from his staid Princeton life was hardly a liberating experience. Rather, he found moving his family around the country to take on various tasks a burden.

In general, Wheeler had little inclination to be freewheeling and zany. Compared to Feynman's roughness, he had the demeanor of a Sunday school teacher. His humor was dry and subtle, rather than outrageous, relying on wit and playful teasing. In short, while the project years offered Feynman a chance to go a bit wild, Wheeler had no such opportunity or desire. Keeping his family happy during all the moves was his priority.

The Wheeler family's whirlwind trip began in the summer of 1942, shortly after their youngest, Alison, was born. John found a rental house in Chicago, close to the university, where they could stay temporarily. As their son, Jamie, was sick with rheumatoid fever, and Janette was recovering from postpartum complications, it was not a fun time.

In March 1943, Compton reassigned Wheeler to a position in Wilmington, Delaware, so that he could work as liaison to DuPont's laboratories. The family was forced to move again. They stayed there little more than a year, before ricocheting, like an electron in a Compton scattering experiment, to the opposite site of the country. From summer 1944 until the end of the war in summer 1945, they resided in Richland, Washington, so that Wheeler could work in neighboring Hanford to supervise the building and operations of a plutonium-production reactor. He still had to "commute" back to Wilmington on occasion—an arduous train ride involving multiple connections.

As part of his Hanford role, Wheeler visited Los Alamos several times to consult with its experts about plutonium safety and other issues. No one was certain how much plutonium could safely be assembled. Would they have to build expensive new storage facilities to divide up the plutonium produced to prevent it from going critical and causing a nuclear meltdown? Or could they reliably keep it in a single container?

Feynman found himself, once again, applying his computational skills to Wheeler's questions. Only now, rather than abstract theory, the work was a matter of life or death. He crunched the numbers and offered Wheeler some data, which Wheeler brought back to Hanford.

After discussing the results with the plant manager, Bill Mackey, Wheeler and his Hanford colleagues devised ways to minimize the risk of accidents.

Wheeler profoundly sympathized with Feynman's anguish over Arline's dire situation. During one visit to Los Alamos he stopped by her hospital to offer his support and wishes for her good health. Feynman was grateful for the benevolent gesture, which fit his image of Wheeler as tantamount to a saint. A saint with wacky ideas, but a saint nonetheless.

With the nuclear effort so much on their minds, Wheeler and Feynman had little time to discuss their former research. At that point, Wheeler probably thought about it more than Feynman did. Feynman calculated; Wheeler dreamed. Even when he tried to focus on more practical matters, Wheeler couldn't help but wonder about nature's hidden truth.

Among the stress of his war work, Wheeler longed for a future when he could once again stare quietly out of his Palmer Lab window, over the vernal Princeton campus, and think deeply about the essence of reality. Could he explain everything in terms of electrons? Are electrons—or perhaps even a single electron—the building blocks of reality? Or might there exist something even more fundamental? Might there be an explanation compelling enough to finally bring the two greatest influences in his life—his mentor, Niels Bohr, and his friend Albert Einstein—into agreement?

EXILED AND EXALTED

By then Bohr had far more pressing concerns than his debates with Einstein over quantum theory. Since April 1940, the Nazis had occupied Denmark. Wanting to continue to lead the Danish scientific community and preserve the integrity of his institute, Bohr had stayed in Copenhagen as long as possible.

In September 1943, the Danish resistance passed along word that the Nazi regime was on the verge of arresting him because his mother was Jewish. Warning him to escape, they arranged passage to neutral Sweden for him and his family. Leaving in the dead of night, clutching what he thought was a green bottle of heavy water (useful to moderate nuclear reactions), Bohr fled with his family across the Danish Strait in a fishing boat.

To place even more distance between the Nazis and himself, Bohr took up an offer to travel to Great Britain by plane. To avoid detection and being shot down over occupied Norway, he was transported via Mosquito, a type of military aircraft capable of flying at high altitudes. As he made his way inside, he was told to don an oxygen mask. His son Aage, who was studying nuclear physics, took a different plane.

For some reason—possibly because his head was too large or he simply forgot—Bohr never put on the oxygen mask and passed out mid-flight. Luckily, he was revived upon landing in the United Kingdom. However, the green bottle that he held onto so firmly turned out to be beer; he had taken the wrong one. The Danish resistance had to sneak back into his house to retrieve the precious heavy water.

In December 1943, Niels and Aage were invited to visit the United States to help support the Manhattan Project. Arriving in Washington, DC, they met with General Groves, who filled them in on the progress so far. To reduce the risk of their being identified by foreign agents, they received the code names Nicholas and James Baker. As the project reached its final stages, they paid several visits to Los Alamos—partly to offer advice about nuclear bomb design but mainly to boost morale.

Feynman recalled the excitement when "Baker and son" first arrived at the lab. From the highest officials on down, everyone was on pins and needles, bracing for the Danish physicist's assessment. "Even to the big shot guys, Bohr was the great God," Feynman remembered.

Much to his astonishment, shortly before one of the Bohrs' visits, Feynman received a call from Aage asking for an early-morning meeting to discuss bomb design modifications. Feynman couldn't imagine why the Bohrs had chosen him, of all people, rather than someone better known. Of course he accepted.

In a secluded office, the three men sat down. Bohr turned to Feynman and began to speak very quietly. "Mumble, mumble, mumble," he said, or something to that effect. He took a puff on his pipe and mumbled again. Feynman had little idea what the notoriously soft-voiced Bohr was talking about. Fortunately, Aage could serve as "interpreter" and explain his father's scheme.

Even in the case of someone as famous as Bohr, Feynman could not help but offer a candid reaction. He explained the impracticality of the technical details of Bohr's suggestions. Bohr appreciated honesty. Although Feynman disagreed with him, Bohr was grateful for the frank assessment.

By mid-1944, the project was in full gear. At the same time, global events had begun to move quickly. On June 6, western Allied troops crossed the English Channel and landed in Normandy, France. D-day, as that pivotal date became known, marked a critical turning point for the European phase of the war. Over the course of months, more and more western Allied troops took advantage of that foothold and marched on to Berlin. Soviet troops, in the meanwhile, marched westward toward the same destination. On Victory in Europe Day, May 5, 1945, Berlin fell to the Allies, and the European phase of the conflict was over. Because the Japanese leadership refused to surrender unconditionally as the Allies demanded, however, the war continued in the Pacific.

Bohr couldn't wait to return to Copenhagen and resume his leadership of the Institute for Theoretical Physics. An important anniversary was coming up for him: his sixtieth birthday on October 7, 1945. Given his pivotal role in modern physics, the focus of enormous admiration and respect, preparations for the celebration would start much earlier.

A hallowed tradition in the physics community is the creation of a Festschrift—essentially a collection of papers by colleagues and former students—for anyone who has made major contributions and is turning sixty. Typically, such a volume pays homage by offering new results and insights based on the honoree's own contributions to the discipline. Given that Bohr's work touched on so many fields, assembling such a tribute to him would not be difficult.

In spring 1945, the prestigious journal *Reviews of Modern Physics* dedicated an entire issue to Bohr that served effectively as a Festschrift. The contributors constituted a who's who of quantum and nuclear physicists, including Max Born, Paul Dirac, George Gamow, and even Einstein, who, along with assistant Ernst Straus, wrote about gravitation and the expanding universe.

A lead article by Wolfgang Pauli about Bohr's legacy helped set the tone. The often cantankerous theorist praised the father of atomic theory effusively. Following Pauli's capstone, article after article in the volume lauded Bohr and called attention to the impact of his contributions.

Despite being bogged down with war duties, Wheeler heartily wanted to submit a piece to the Bohr celebration. His results with Feynman seemed the perfect topic, as he had felt uneasy setting them aside for so long. In theoretical physics, delaying publication of a finding risks that someone else will come up with the same idea and publish first. The tribute issue, Wheeler thought, offered the ideal opportunity to bring to the physics community's attention at least some of the action-at-a-distance results. He quickly assembled a paper—mostly written by him but crediting Feynman as coauthor—reporting a large chunk of their findings.

WHEN THE FUTURE SHAPES THE PAST

Wheeler and Feynman's inaugural published work (aside from a conference report), "Interaction with the Absorber as the Mechanism of Radiation," begins rather unusually with extensive footnotes attached to its title. In them, Wheeler explains that they had done much of the work for the paper several years earlier, but the war had delayed completion of the project. Therefore, he notes, readers should view the article as a work in progress—just one installment in an anticipated multipart series.

Feynman deferred to Wheeler's judgment on the paper's organization and content. However, he personally viewed the "multipart" aspect as unnecessarily ambitious and verbose. As it turned out, they would end up writing just one more paper together on the topic—with a four-year gap between the two works.

As befits its placement in a commemoration of Bohr, an enigmatic quote by the great Danish physicist, extracted from a book he had written in 1934, tops off the article: "We must, therefore, be prepared to find that further advance into this region will require a still more extensive renunciation of features we demand of the space time mode of description."

Wheeler liked to ground his hypotheses, no matter how lofty and ethereal, in the concrete foundations set down by his respected predecessors and mentors—especially Bohr and Einstein. For him, invoking either assured readers that his "crazy ideas" didn't emerge out of the blue; rather they sprang from deep contemplation about extending more mainstream work. Consequently, Wheeler likely included Bohr's quote to demonstrate the Danish thinker's open-mindedness toward recasting quantum physics on a different footing—one with an alternative view of space and time. By implication, the absorber theory, with its symmetry between forward and backward motion in time, offered the next step forward in the progression of quantum theory.

The paper's introduction continues the "sales pitch" by emphasizing the tradition of action at a distance in physics, as formulated by Isaac Newton, German mathematician Carl Gauss, and others. While these ideas had been abandoned in favor of energy fields filling all of space, perhaps it was time to resurrect them to resolve certain conundrums regarding radiation. In particular, the article continues, reviving action at a distance would offer a superior way of explaining the phenomenon of radiation resistance.

Wheeler then calls upon another of his key mentors, Einstein, to help justify the absorber idea. He mentions how he had learned from Einstein about the work of obscure Dutch physicist Hugo Tetrode, who had speculated in 1922 that any radiation must not only have a source but also something to absorb it. As Tetrode had put it, "The Sun would not radiate if it were alone in space and no other bodies could absorb its radiation."

If Feynman had written the paper on his own, he would have left out most of the background, the speculation, and the references. He hadn't even read Tetrode's work and had little interest in doing so. Rather, he would have cut to the chase—simply laying out the physical problem of radiation resistance and presenting his results. His version would have been more like his PhD thesis, succinct and direct. Ceding most of the writing to Wheeler had resulted in a very different paper.

After detailed calculations of how radiation behaves in the Wheeler-Feynman absorber theory—the crux of Feynman's work—the paper reverts to Wheeler's passion for speculative notions. It concludes on a rather philosophical note: a discussion of time's arrow. If an absorber's action in the future could determine a radiating particle's behavior in the past, what does that tell us about causality? As Wheeler writes,

> Pre-acceleration and the force of radiative reaction which calls it forth are both departures from that view of nature for which one once hoped, in which the movement of a particle at a given instant would be completely determined by the motions of all other particles at earlier moments. . . . The past was considered to be completely independent of the future. This idealization is no longer valid when we have a particle commencing to move in anticipation of the retarded fields which have yet to reach it from surrounding charges.

Wheeler and Feynman's paper thus ends on a revolutionary note: imagining a world in which the future influences the past, and vice versa. By removing any distinction between forward and backward causality, it goes well beyond stating that the particle realm simply appears reversible in time. Rather, it makes the future and past equally relevant to the future.

The difference between apparent and actual time reversibility is important. Many things look time reversible without actually being so. If you yelled "ouch," ran forward into a wall, bounced off it, ran backward, and screamed "ouch" again, a video of your antics might look symmetric in time. However, you couldn't rightly claim that the first "ouch" was due to your hitting the wall, before that actually happened. In the Wheeler-Feynman absorber theory, in contrast, particles feel the impact of events taking place in the future—the advanced signals generated by the absorption of their radiation.

From the seeds of Wheeler and Feynman's concept of advanced signals, an entirely novel approach to electrodynamics would sprout. Coinciding with the new approach was a revolution in thinking about time. Their scheme unshackled particle physics from the fetters of time's forward motion, allowing the past, present, and future to speak with each other.

UNLEASHING THE DEMON

The explosion of an atomic bomb relies on a cascade of nuclear reactions, each of which, in principle, might be reversible in time. If, as the Wheeler-Feynman absorber theory suggests, the mechanisms of radiation are time symmetric, one might imagine videotaping each of the particle interactions involved and running that film backward in time. The mushroom cloud would retreat, the uranium and other materials would reassemble, and the device would restore itself to pristine condition.

In terms of time's arrow, large-scale reality is nothing like the particle world, however. For one thing, the second law of thermodynamics applies, mandating that for many processes a large chunk of usable energy transforms into waste energy. The massive blasts of heat and widespread destruction caused by atomic explosions offer prime examples of such irreversibility. No one could begin to imagine how to turn back the clock and undo the devastating impact of a nuclear bomb dropped on a city.

After the bombings of Hiroshima and Nagasaki in August 1945, Japan surrendered, World War II ended, and the Allies celebrated long-awaited victory. Harry Truman, who had assumed the US presidency after the death of Franklin Roosevelt, made the decision. Amid relief that the conflict was finally over, many questions arose about the ethics of incinerating and irradiating large civilian populations—even for the ostensible purpose of avoiding more casualties. Many of those who had warned Roosevelt about the Nazi efforts, such as Einstein and Leo Szilard, or participated in the Manhattan Project itself, such as Wilson, were horrified that the United States had dropped the bomb on Japanese civilians when these scientists had intended their work to serve only as a hedge against possible German efforts. The fact that the German nuclear program had never truly gotten off the ground cast the actual use of such weapons in a more sinister light.

Might there be a way of putting the demon back in the bottle and preventing a future conflagration? Once revealed, the secrets of building nuclear weapons could never be "unlearned." Bohr, a firm supporter of openness in science, argued vehemently that all the information should be made public. Along with Einstein, Wilson, Szilard, and others, he advocated for international control of nuclear weapons. A large fraction of the physics community, including many who had worked on the bomb, became supporters and organizers of arms control organizations.

Others, notably Teller, warned of an arms race with the Soviet Union and urged the Western powers to keep a lid on nuclear secrets.

He urged development of a "superbomb," later called an H-bomb or hydrogen bomb, to maintain an edge over enemies of democracy. He would continue to pursue advanced weapons research, long after the Manhattan Project was over, and become known as the "father of the H-bomb." Famously the film *Dr. Strangelove, or: How I Learned to Stop Worrying and Love the Bomb* satirized him and those taking his hawkish stance.

SAVING JOE

Though much less vocal and militant about his decision than Teller, Wheeler would surprisingly follow a similarly pro-nuclear path. He would continue his weapons research up until 1953, becoming an important contributor to and advocate of H-bomb development.

Why would Wheeler, a quiet, peaceful disciple of internationalists Bohr and Einstein, pursue such a hawkish course? A tragedy—the battlefield death of his younger brother Joe, only thirty at the time—may have steered him toward supporting more nuclear weapons research and development.

Joe was a talented young historian who had received his PhD from Brown University. With a promising career ahead of him, he enlisted as a private first class in the US Army. Fighting Germans in Italy, he found himself in the thick of battle. Sometime in 1944, he sent John a postcard with a simple but powerful message: "Hurry up!"

In hindsight, Wheeler surmised that his brother, who knew about his fission research, suspected that he was developing a superweapon with the potential to end the war. While the Allies had Hitler in their crosshairs, a powerful show of force would help speed his defeat along.

Tragically, shortly after Joe sent the card, he went missing in action; his decomposed body wasn't found until 1946. When John and his family learned the bitter truth, the loss was devastating.

From that point on, thinking over and over again about his brother's message, Wheeler became obsessed with an alternative history in which he had never put aside his fission research. In the parallel, early 1940s of his imagination, rather than switching gears and focusing exclusively on his electrodynamics work with Feynman (along with his teaching and other duties), he would push strongly for the development of nuclear weapons. Lending his considerable research and organizational skills to an expedited Manhattan Project, he'd make sure that the Roosevelt administration devoted enough manpower and resources to the mission. By the time the United States entered the war, the bomb would already be well under development. If it were ready by mid-1944, Wheeler envisioned, perhaps Nazi Germany would surrender one year earlier, sparing millions of lives—including numerous soldiers and civilians caught up in the final stages of the conflict and a large portion of those murdered in the Holocaust.

"I am convinced that the United States, with the help of its British and Canadian allies, could have had an atomic bomb sooner . . . if scientific and political leaders had committed themselves to the task earlier," Wheeler later wrote in his memoirs. "One cannot escape the conclusion that an atomic bomb program started a year earlier and concluded a year sooner would have spared 15 million lives, my brother Joe's among them."

Decades after his brother's death, Wheeler would mention, in public talks about the era, his guilt over not pushing earlier and harder to develop the atomic bomb. Tears would well up in his eyes, as he described what could have been if the Allies had defeated the Nazis sooner and his brother were still alive. Once again, those heartbreaking two words: "What if?"

DOOMSDAY CLOCK

A telling symbol of the atomic age is the "Doomsday Clock," which began appearing in 1947 on the cover of the *Bulletin of the Atomic Scientists*—a magazine started by Manhattan Project veterans who favored arms control. The clock warned of the imminent threat of nuclear disaster; its hands, initially set at seven minutes to midnight, have been moved forward or backward since to represent the waxing and waning of such danger. Midnight symbolizes nuclear catastrophe—increasingly associated, as bombs have become more powerful, with the end of civilization and perhaps even life on Earth.

Fear of apocalypse pervaded the very first test of an atomic bomb, dubbed by Oppy the "Trinity test." It took place in the early morning of July 16, 1945, in the desert region called Jornada del Muerto (Journey of Death), about two hundred miles south of Los Alamos. A plutonium bomb was placed in a firing tower several days before the countdown to detonation. Bethe and Feynman had carefully calculated the amount of energy they expected the bomb to release—through what became known as the "Bethe-Feynman formula." To check if their estimates and other predicted aspects of the explosion were on the mark, instrumentation had been set up around the testing site. Scientific observers and major officials were assigned to bunkers about six miles away from the tower. Other witnesses to the event, including Feynman, remained some twenty miles away. Each was allocated special dark goggles for eye protection. As countdown neared, nervousness set in about the bomb's impact. While some, such as General Groves, worried that it wouldn't work at all, others feared that it might be too effective and set off a chain reaction of devastating events, perhaps even igniting the atmosphere. With gallows humor, Fermi took bets on whether such atmospheric incineration would happen, and if it did, whether it would wipe out just New Mexico or the entire world.

Perpetually curious, Feynman decided to skip the dark goggles, figuring they'd obscure his view, and watched the explosion from inside a weapons carrier truck. Its windshield, he reckoned, would protect his eyes from any ultraviolet radiation emitted by the bomb, while letting visible light through. He braced to see if his calculations were right and the device actually worked.

Suddenly his eyes registered a blinding white flash. Success! Instinctively, he turned his head away and saw a purple splotch. No matter where he looked inside the truck, he saw purple. Even with his eyes closed, he saw the blotch. Rather than panicking, he reassured himself that he was just viewing a temporary afterimage. So he opened his eyes again and glimpsed a yellow ball growing in the distance like a second sun being born. Spreading even more, it darkened into orange. The ball rose on a thick stem of smoke, resembling something like a mushroom. All the while, a calm, internal "commentator" gave him a blow-by-blow report of the physical laws underlying each step of the process. Finally, he heard a thunderous boom—sound, he knew, takes much longer to arrive than light. At twenty miles away, the difference was a minute and a half.

Several weeks later, when two atomic bombs were dropped on Japan, he felt no horror or remorse. He shared the general sense of relief that the war was over. It didn't occur to him to feel guilty or worried. Back at the lab, Wilson, who had pushed him into joining the project, stunned him by decrying the evil thing they'd just brought into the world. Why would Bob, he wondered, disavow his own baby? For Feynman, only much later did the implications of the technology he had played such a major role in developing begin to sink in.

Why was a man as sensitive as Feynman initially so nonchalant about the nuclear explosions and their destructive power? Perhaps it was because, weeks earlier, his world had already ended. His true love was gone forever, and his heart was as barren as the Trinity site.

In mid-June Feynman had received a distressing call from Arline's father, who was visiting her at the hospital, alerting him that her condition was dire. Borrowing Fuchs's car, he rushed to Albuquerque as fast as he could, contending along the way with a flat tire two different times. When he arrived, she was barely conscious and struggling for breath. She died on June 16.

All the while, Feynman remained surprisingly dispassionate. His brain told him that dying was a natural physiological process. He and Putzie had reveled in so many fun adventures together, even when she was ill. Whether they had spent many decades or only a few years enjoying married life, it was all the same in the end.

Feynman noted that the clock in Arline's hospital room was strangely frozen at the time of her death: 9:21 p.m. Did time really stop when his sweetheart left the world? Of course not. His ever-churning mental gears offered a much more reasonable explanation. The clock was faulty, and the nurse had knocked it over when checking the time of death.

A clock had stopped in Feynman's emotional life too—although he didn't fully realize it at the time. To his friends back at Los Alamos, he presented the same carefree attitude he always had. Dispassionately he watched the Trinity test and learned about the bomb drops on Japan. He celebrated along with the rest. But gradually he began to realize that something was terribly wrong. Something essential was missing. His body and brain were on automatic pilot; the central processor had shut down, and the output was meaningless. What mechanic could fix his broken inner life and make it tick again?

CHAPTER FOUR

THE HIDDEN PATHS OF GHOSTS

"At every fork in the road, we took the road in the worse condition, the one that looked interesting to us."

—Michelle Feynman about her father, Richard, in *Perfectly Reasonable Deviations*, xix

With the end of World War II, great sadness about the massive loss of life tempered public jubilation. Horrific devastation throughout much of Europe, Japan, and other parts of the world reminded survivors every day that history had taken a terrible course. Speculation abounded about what the world might have been like if Adolf Hitler had never existed. Would such an alternative reality have been more peaceful—or were war's terrors in some way inevitable? Might other leaders, just as brutal, have taken his place?

On the flip side, many pondered what would have happened if Hitler had won. Alternative history scenarios abounded, exemplified by Philip K. Dick's classic novel *The Man in the High Castle* (published in 1962 and dramatized as an Amazon television series beginning in 2015). In that work, a series of circumstances leads to an Axis victory over the Allies. Nazi Germany and Imperial Japan divvy up the planet—save for certain neutral zones.

Dick's vision drew on several sources, including Ward Moore's 1953 novel *Bring the Jubilee*, in which the South won the American Civil War, and the ancient Chinese *I Ching* (Book of Changes). The latter has served for centuries as a divination system in which randomly selected symbolic hexagrams, each with a distinct pattern of line segments, portends one of many possible futures.

In Dick's chronicle, several characters rely on the *I Ching* in their decision making. One of them is Hawthorne Abendsen, a writer who uses it to construct a novel-within-a-novel titled *The Grasshopper Lies Heavy*. Forbidden in the Nazi-controlled regions but popular elsewhere, it describes yet another variation of history in which the Nazis are defeated in a different manner and time frame than actually happened.

Another *I Ching* devotee, Juliana, a fan of *Grasshopper*, seeks out Abendsen to discuss the meaning of his novel. After meeting him at his house, she reaches a startling conclusion. The fictional Nazi defeat, she discovers, is truer than the timeline in which she lives. In other words, the Nazis were meant to lose, not to win. She realizes that the chronology she knows is really a mirage: a false path through time.

Dick's convoluted, speculative tale revisits a question asked by many philosophers: Could other narratives of history lie on unseen pages of an immense digest of possibilities? If so, might it be possible to sense the unrealized chronologies, as several characters do in the novel?

German mathematician and logician Gottfried Leibniz speculated that God has access to every alternative. After weighing all of them, he chooses the optimal path. In the case of World War II, presumably that would be the actual scenario: the Nazis' defeat in 1945. If Hitler could have been thwarted earlier and millions of lives saved, Leibniz's notion suggests, God would have chosen that path.

In *Candide*, the French writer Voltaire famously satirized the Leibnizian notion that we live in the "best of all possible worlds." No matter what horrors arise, a character, Dr. Pangloss, remains optimistic and convinced that the state of affairs was meant to be. Otherwise, God would have selected a better outcome.

While the idea of alternative histories remains incredibly speculative in any discussion of human events, Richard Feynman's path integral method, which John Wheeler dubbed "sum over histories," demonstrated that the notion applies to the quantum world. Calculating the results of any particle interaction must take into account a weighted tally of every possible chronology. In manipulating that abacus of possibilities, the classical path turns out to be the optimal one: the best of all possible worlds. It is as if nature has its own *I Ching*.

POSTWAR BLUES

Though Wheeler and Feynman were relieved that the conflict was over, for both men the immediate postwar period—particularly 1946—was a somber time. The loss for Wheeler of his dear brother Joe, and for Feynman of his beloved wife and muse, Arline, cast dark shadows over each. Wheeler felt terrible that his young protégé was now a widower.

For the Wheelers, a long-awaited homecoming tempered their wartime tragedy. When they returned to their comfortable house on Battle Road in Princeton, their vagabond years seemed finally over. (They'd travel to Los Alamos several years later, but the journey wasn't quite as frantic.) Focusing on their children's schooling and other domestic matters helped them put the war years behind them.

Feynman had no such family life to distract him from brooding. The project to which he had devoted so much energy for the past

The Wheeler family house on Battle Road, Princeton, New Jersey (*Source:* Photo by Paul Halpern).

few years had vanished in clouds of radioactive dust. Moral issues aside, he had been immensely proud of his successes at Los Alamos—including the safecracking and other antics. All that was over, leaving an immense void.

Adding to Feynman's pain, in October of that year his father died. As someone with an unlimited passion for scientific curiosity, Melville Feynman had always motivated Richard, who often thought about how to explain ideas to him. With his dad gone, he worried about how his mother would cope and became very protective.

One afternoon, Feynman was meeting his mom for lunch in New York City, when a wave of depression came over him unexpectedly. Observing the street life around him—all the businessmen, tourists, and others roaming the skyscraper canyons of the city—he mentally calculated how many blocks an atomic bomb would decimate. He thought about the feasibility of making destructive nuclear weapons and trembled at the possibility that Manhattan could end up like Hiroshima and Nagasaki. Suddenly he realized the extent of the horror that he and his colleagues at Los Alamos had unleashed. There was no hope for the world, he concluded. Everything was futile.

As a human, not an elementary particle, like a positron, or an advanced wave, Feynman knew that he couldn't go back in time and change history. Rather, he would learn an important lesson from his mistake. He had erred in assuming that the initial mission of the Manhattan Project—to prevent the Nazis from developing the bomb and gaining a monopoly on its use—justified its completion. On realizing that the Nazis had made no progress, he and others should have opted out. The world would have been better off without the sword of Damocles of nuclear war hanging over it. For future missions, he vowed, he would constantly reexamine his assumptions and adjust his plans to changing conditions.

Around the same time, Feynman felt a powerful urge to comfort Arline and tell her how much he still loved and missed her. He decided to write her a letter saying the things he wished he could have told her when she was still alive. Recalling her desire not to burden him with her problems, he explained how, on the contrary, she had helped him so much, even during her time of illness. She was his inspiration. Without her, life simply wasn't as fun. "You were the 'idea-woman' and general instigator of all our wild adventures," Feynman wrote.

Feynman understood well the paradoxical situation of writing to someone deceased. For one thing, as he pointed out, he obviously couldn't mail the letter to her. Yet he confessed that Arline still meant more to him than any living person. "I love my wife. My wife is dead," he dolefully concluded. The letter, never sent, was well worn—indicating that he likely read it again and again.

HOLY SIMPLICITY

Wheeler's return to Princeton after the war freed him to return to his twin intellectual passions: teaching and fundamental research. Once again, he could gaze out his window in Palmer Laboratory (where his office had been relocated from Fine Hall), contemplate the twisty patterns of the trees, and meditate deeply about the true essence of nature.

Philosophically, Wheeler continued to believe that our complex world might be built of simple components—like an intricate model of a city erected using basic Lego blocks. His research with Feynman seemed to confirm his hunch that everything stemmed from electrons, along with their positive counterparts, positrons. There might be a single electron, zigzagging backward and forward in time, or multiple electrons and positrons.

Like musical duets filling the air with beautiful harmonies, electrons create rainbows, sunsets, thunderbolts, and so many other types of visible and invisible luminous displays through their mutual interactions. While standard quantum theory asserted that photons serve as intermediaries—what became known as "exchange particles" that convey force when they are swapped back and forth—the Wheeler-Feynman absorber theory removed that requirement. Electrons (and positrons) could generate light through their own harmonizing. The particle world thus became simpler.

Wheeler attributed his zeal for minimalism to his austere Protestant background. Specifically, he was a lifelong Unitarian, a faith that distills unifying principles from diverse beliefs. In line with such values, he ardently sought the essence of things, looking beyond superficial differences.

Like a Shaker furniture maker pursuing his craft with saw, hammer, and nails, Wheeler deemed simplicity a virtue. His construction materials would change over time, but his pursuit of the basics would remain constant. He described his passion as follows:

> There grew up in time, to be sure, a litany that many a student was taught to repeat with a mindless faith of a catechism: there are four forces: the strong force, the weak force, the electric force, and the gravitational force. But my Protestant upbringing made me reject this catechism. What simpler faith could I put in its place? Ideals of unity and simplicity, unattainable now and perhaps for years to come. Take one force, electromagnetism, and explore and exploit it to the limit. That made a program pure enough and ambitious enough that I could give myself to it wholeheartedly.

To continue his simplicity mission, Wheeler aspired at that point to construct all the other known particles and forces from electrons and positrons. He published a paper in which he coined the term "polyelectrons" to describe kinds of "atoms" and "molecules" built from electron-positron pairs. Somehow he hoped to identify such constructs with the inhabitants of the particle kingdom. The pairing of a single electron with a single positron in an extremely unstable mock atom became known as a "positronium." Two positronium atoms would form a molecule of dipositronium. Wheeler's innovation was imagining a world built up from electrons and positrons alone, grouped into atoms and molecules, that would somehow be more fundamental than the standard mix of protons, neutrons, and electrons. While ultimately his vision of building protons and neutrons from positrons and electrons proved impossible, he was on the right track. Decades later, scientists learned that protons and neutrons each consist of quarks and antiquarks—which, as point particles, are the "cousins" of electrons and positrons. So the particle family turned out to be larger than he thought.

Wheeler's polyelectron research, though speculative, won him acclaim. The New York Academy of Sciences, an esteemed group of scholars, honored him in 1947 with its prestigious A. Cressy Morrison Prize and published his article on that topic in its annals. It was the first of numerous honors and awards Wheeler would receive in his long career.

By that time, realization was growing that nature had at least three—and perhaps four—fundamental forces. Theorists sought to understand these mechanisms through the language of quantum mechanics. Wheeler and Feynman had focused on electromagnetism—long understood classically, via Maxwell's equations, and ripe for a quantum description. Another well-studied interaction was gravitation. Albert Einstein's general theory of relativity had proven a comprehensive, classical way of understanding gravity. A few theorists had tried, unsuccessfully, to quantize that force as well.

While theories of electromagnetism and gravitation explained an incredibly broad range of natural phenomena—from the whirling of motors to the revolving of planets—they could not address several distinct kinds of behavior on the nuclear level. The first was the radioactive decay of neutrons into protons, electrons, and (what became known as) antineutrinos. That process, called "beta decay," lacked a full explanation—despite attempts by Enrico Fermi and others to model it theoretically. The complete description, which Feynman would play a major role in developing, came to be known as the "weak interaction."

Another important nuclear process requiring a satisfying explanation was the force gluing together nucleons (protons and neutrons) within atomic nuclei. This powerful but short-ranged attraction was later dubbed the "strong interaction." Hideki Yukawa had proposed in 1935 a possible mechanism involving an exchange particle called a "mesotron"—later changed to "meson." Unlike photons, mesons would have significant mass. As with heavy bowling balls that could be tossed only a brief distance into the air, their heft would significantly shorten their range. Hence the strong nuclear force would be confined to subatomic spans.

By odd coincidence, the following year, researchers Carl Anderson and Seth Neddermeyer analyzed a stream of cosmic radiation and discovered a particle roughly matching Yukawa's predicted mass. It became known as the "mu meson," later shortened to "muon." Unfortunately, physicists came to realize that the mu meson didn't quite achieve the task set by Yukawa. In fact, if recruited to bind nucleons together, it would have immediately been fired for incompetency. Muons don't even experience the strong force. They don't seem to play any special role in nature, aside from their existence in cosmic rays and production in various processes. Addressing muons' lack of purpose, physicist I. I. Rabi famously asked, "Who ordered that?"

The proper Yukawa particle, called the "pi meson," or "pion," would be found in 1947 through another analysis of cosmic rays. Pions are short-ranged, massive particles that respond to the strong force and thereby fit the bill. Decades later, however, physicists would realize that they aren't fundamental. The true mechanism of the strong interaction involves "gluons," another type of exchange particle.

Wheeler hoped that polyelectrons would help model mesons and other particles found in cosmic rays. By constructing mesons from the building blocks of electron-positron pairs, he hoped to explain the abundance of exotic particles and forces using familiar electromagnetism and the commonplace electron. Just as the periodic table shows how chemical elements consist of nuclei and electrons, perhaps arrangements of polyelectrons, he thought, could do the same for elementary particles. On the discovery of pions, Wheeler would tell the *New York Times*, "More and more the possibility suggests itself to us that all the heavier particles are built up in some way we do not understand out of positive and negative electrons."

Wheeler was never one to bet everything on a single horse, however. Wagering his scientific reputation on polyelectrons, an unproven, speculative hypothesis, would not be wise. Throughout the late 1940s and early 1950s he published a number of more conventional papers in nuclear and particle physics—including descriptions of muon and pion interactions, discussions about the origin of cosmic rays, and an analysis of certain processes in which two photons are emitted.

As a respected heir to Niels Bohr's legacy, including their codevelopment of key work on nuclear fission that helped make the bomb possible, Wheeler was now very much in demand to give public lectures as well as serve on government committees. He published several articles on the future of nuclear power, a topic in which he had special expertise. He also contributed biographical pieces about Bohr and the pioneering American physicist Joseph Henry—signs of his burgeoning interest in the history of physics.

In September 1946, Bohr returned to Princeton for a conference titled "The Future of Nuclear Science," coinciding with the bicentennial of the university's founding. Wheeler was delighted to host his mentor, along with many other notable physicists—including Feynman, Rabi, Fermi, Robert Oppenheimer, and Paul Dirac—to discuss the postwar direction of physics. It gave Feynman a chance to chat briefly with Dirac about his way of applying the least action principle to quantum mechanics, which was, after all, an extension of Dirac's work. Dirac listened for a moment but was focused on his own ideas.

Feynman sounded positively Wheeleresque in his reported comments at the conference about the need for simplicity in particle physics: "What will the fundamental particles turn out to be—more particles or less particles? Or perhaps all our so-called 'different' particles are not 'different' particles but different states of the same particle. . . . We need an intuitive leap at the mathematical formalism, just as we had in the Dirac electron theory; we need a stroke of genius."

One topic of discussion at the conference was the flood of government and industrial money into science and its possible corrupting effects. Many participants argued that physicists needed to fight for their independence.

Wheeler firmly believed in a citizen's duty to support his government, both in peace and war. He kept closely in touch with his former Manhattan Project colleagues, vowing that as a nuclear scientist he must keep abreast of military as well as civilian developments. He had pledged not to make the same mistake (in his view) of setting aside weapons research and leaving decision making to politicians. Physicists, he determined, must remain active in the nation's defense to prevent ill-informed choices.

Feynman, on the other hand, had completely lost his taste for military work. While not politically active like Robert Wilson, he had largely followed Wilson's path. Wilson had set aside fundamental research and started military work for the express purpose of stopping Hitler. Once he learned that the Nazis didn't have the bomb, his interest in weapons work had evaporated. While he persisted until the project's completion, his heart was not in it.

Following the war, Wilson couldn't wait to return to a civilian life exploring the wonders of the physical world. Feynman felt the same. While he didn't condemn bomb development, he had no desire to resume it. Explaining that he had other plans, he politely refused various invitations to consult at Los Alamos and elsewhere. Decoding nature's enigmas was a far more rewarding pursuit than finding new ways to blow it up.

BETHE'S BOYS

After the Manhattan Project, Feynman had a variety of career options. A number of universities would have been happy to have him on their faculties. He could have tried to reactivate the offer to work at the University of Wisconsin. He might have pursued Berkeley, where Oppenheimer had been pushing for him. Raymond Birge, the department chair, had been slow in making an offer but eventually did.

However, he found Hans Bethe's offer to teach and do research at Cornell the most tempting. Feynman respected Cornell's accomplished nuclear experimental group and thought that its results might stimulate him to craft better theory. On a practical note, it was within a few hours of New York City, allowing him to reconnect with his family and attend some of the national physics conferences held there.

Another option that Feynman ended up turning down was becoming a researcher at the Institute for Advanced Study. There, he wouldn't have to teach at all. He could simply ponder the ultimate questions to his heart's content while earning a generous salary.

Feynman had seen how Einstein and others isolated at the institute had lost their grounding in reality. Current developments in physics sometimes never penetrated the institute's walls. Without interactions with students, its members' minds were free to wander, to wonder, to ruminate—but about what? Feynman had no interest in abstract or purely mathematical questions. Nor, at that point, did he want to tackle the question of unifying the forces of nature. Without the impetus of preparing and teaching classes, he would be less motivated to keep up with the times. Rather, like Einstein, he might become enmeshed in a theoretical dream world.

Moreover, Feynman knew that for any theorist, the road to solving problems could be treacherous. If he was at a teaching institution, whenever he got stuck in a rut, he could switch his focus to pedagogy. This would help insulate him against feelings of worthlessness. After all, he loved teaching and found it easy. Deeply depressed at that time and finding it hard to even think about research, he thought such a path the safest. Consequently, he chose Cornell above the alternatives.

Feynman's first day on campus was in some ways akin to the biblical Joseph and Mary's arrival in Bethlehem: there was no room in the inn. Arriving after midnight, without reservations, he hadn't anticipated that all the rooms in all the hotels would be booked. In Ithaca, the compact, hilly college town where Cornell makes its home, the start of the semester is always a busy time. Nonetheless, Feynman had simple needs. He looked around for a building that happened to be unlocked, found a quiet space with a sofa, and fell asleep right there.

When he woke up in the morning he reported to the physics department and asked where his class on mathematical methods of physics would be held. Told that he was a week early, he groaned that he had prepared his first lecture and was dismayed to have to wait so long to deliver it. With no leads on lodging, he headed back to the building where he had stayed the night before—surreptitiously, he had thought. However, the secret was out. A housing official there cautioned, "Listen, Buddy, the room situation is tough. In fact, it's so tough that, believe it or not, a professor had to sleep in the lobby here last night."

Luckily, many of the physicists at Cornell knew Feynman and his quirks very well. There, Bethe had assembled a youthful cadre of crackerjack whizzes who had served at Los Alamos and other atomic labs. Along with Feynman and Wilson, who was hired soon afterward, recruits from the Manhattan Project included Philip Morrison, who had brought the core of the Trinity bomb to the test site in the back of a sedan and helped load the bombs designated for Japan onto the bomber planes. They formed a consummate team of experts who transferred their formidable skills from bomb design and assembly to tackling some of the most challenging questions in contemporary physics. For those Los Alamos veterans, the sound of Feynman's bongo drumming at all hours was as familiar as the chirping of birds in the spring.

BROKEN HEARTS AND WOBBLING PLATES

Spring is always welcome in places such as Ithaca that get blasted with snow and ice each winter. During the coldest months of the year, when it takes a prayer to tackle the steep, slippery roads, many couples huddle by the fire, happy to bask in the warmth of each other's company. For a young widower like Feynman, those frigid gray days were awfully lonely.

Feynman tried hard to get Arline out of his mind and resume his research. Yet the more he attempted to turn his attention back to theory, the more frustrated he became. He worried more and more that he had lost his touch. Cornell had made a mistake in hiring him, he thought.

Channeling his energy into teaching, he quickly realized that he had found his calling. In the classroom, he was a veritable showman—a master of science entertainment. All eyes were on the maestro as he spun his crazy tales about the weird wonders of nature. In the same way that he impressed his sister, Wheeler's children, and others, he regaled his students with zany demonstrations and colorful explanations of how things work. Just as he threw soup cans in Wheeler's house, he would find anything available—simple, common items—to make his points about the laws of physics and how they affect everything.

After class, however, he'd get back to thinking that he ought to do something else besides teaching. The fact that universities such as Berkeley and think tanks such as the Institute for Advanced Study had made plays for him seemed ridiculous. With almost nothing to show for his research, except his paper with Wheeler, how could they place him in the same category as Oppenheimer and Einstein?

Fortunately, his colleagues at Cornell truly appreciated his talent and were extremely supportive. They put themselves in his shoes and imagined how little work they'd get done if they'd lost their spouses at such a young age. Wilson counseled him not to worry about impressing them. His teaching was great, and research was always a gamble. With Wilson's encouragement, Feynman's cloud of depression finally started to lift.

Anxiety works in strange ways. Sometimes if you keep pushing and pushing toward a goal, you feel as frustrated as Sisyphus watching the rock he has hoisted up a mountainside roll right back down again. Feynman's initial attempts at research after the war were like that. But if you focus on another task, such as teaching, and turn the adverse activity into something like a hobby, suddenly it becomes a cool diversion that you anticipate with glee.

Shortly after Wilson's pep talk, Feynman was sitting in the Cornell cafeteria, when he observed someone tossing a plate in the air, just for a lark. The plate wobbled and spun as it flew. He found that by noting the position of the Cornell insignia, which looked like a red spot swirling around on the plate, he could count the ratio of the rates of its wobbling and its spinning. He soon wrote down the dynamical equations for wobbling, spinning objects and confirmed the identical ratio. He was right! What was the point of such an exercise? As he explained to a perplexed Bethe, who saw little connection to his research, it was fun calculating such things. That cerebral workout helped renew his faith in his own prowess and his passion for physics.

Once his mental gears were churning, he starting thinking about revisiting his thesis topic. His work with Wheeler had made him a master craftsman in assembling models of interacting particles. He had become a virtuoso at wielding the most effective mathematical tools. Yet his thesis was limited because it did not include the effects of special relativity, quantum spin, and other important aspects of modern physics. He thought again about the open questions raised by Dirac. So many aspects of the quantum description of electron behavior rested on shaky foundations. It was time to get back to work; it was time to bring back the fun.

CRACKS IN DIRAC

When Wheeler and Feynman developed the absorber theory, they were far from the only physicists trying to patch cracks in the edifice developed by Dirac. For example, Hendrik "Hans" Kramers, a prominent Dutch physicist, had persistently pointed to what he saw as major gaps in quantum electrodynamics. Kramers had been Bohr's right-hand man for many years, so his influence held considerable sway.

Despite the beauty of Dirac's work, Kramers keenly understood its flaws. He had made a study of all the mathematical inadequacies in the theory. Most of those defects could be summed up (or, rather, not summed up) in a single word: divergence. We've mentioned how an infinite sum in one of Dirac's calculations motivated Feynman's work with Wheeler and led to their absorber theory. Kramers pointed out example after example of calculations in Dirac's version of quantum electrodynamics that diverged. Infinities are dandy in abstract mathematical ruminations, but in physics they are killers. An infinite sum renders any computation nonsensical, especially if meticulously constructed experiments point to a finite value.

Contrary to the Wheeler-Feynman absorber theory, Kramers believed that electrons must interact with electromagnetic fields. In light of the elegance and simplicity of the field explanation in electromagnetism, he did not favor the radical surgery of removing the fields completely. Instead he sought to separate the effects of the fields from the bare electrons themselves. "Bare," in this context, referred to the hypothetical situation in which no fields existed. The experimentally measured mass of an electron, he argued, combined the bare electron mass plus the mass acquired by the electron's interaction with the field it created. (Mass arises from energy, following Einstein's famous relationship.) In a process dubbed "renormalization," Kramers argued, the self-interaction mass could be subtracted from the experimental mass to obtain a more realistic value. Ultimately, Kramers hoped to remove the infinities from quantum electrodynamics to allow for finite, reasonable calculations of the self-energy and other physical quantities.

Naturally, any revolutionary approach called for careful testing to confirm all its predictions. That held for Kramers's hypothesis, the Wheeler-Feynman absorber theory, and all other attempts to resolve the divergence conundrum. As it turned out, precise new experiments conducted in early 1947 and revealed at a pivotal conference on Shelter Island would lead prominent physicists, including Feynman, to use the ideas of Kramers and others to place quantum electrodynamics on firmer ground.

GATHERINGS OF MINDS

The genesis of the Shelter Island conference and several successor meetings was a brilliant idea by Duncan MacInnes of the Rockefeller Institute to channel the scientific energy of the war effort's superstars into resolving deep questions in physics. He proposed the concept to the National Academy of Sciences (NAS), whose officers supported it enthusiastically. The series would have an immeasurable impact on modern science.

NAS president Frank Jewett worked with MacInnes to guide the structure of the conferences. He promoted the notion of short, topical meetings restricted to small groups of leading experts. With Jewett's design in mind, the NAS generously offered ample funding for the retreats, affording comfortable, scenic settings.

After an inaugural conference dealing with biological physics, a second was planned on the topic "fundamental problems of quantum theory." For that, two prominent physicists offered their assistance, Karl Darrow, the vociferous secretary of the American Physical Society, and his friend León Brillouin, who had much experience with the famous Solvay Conferences held in Europe. Brillouin, in turn, suggested asking for Wolfgang Pauli's help.

Pauli's advice was far from what MacInnes and Jewett had in mind. He painted a picture of a large conference, populated with older researchers who had been involved in the quantum enterprise before the war and wanted to resume their discussions. Pauli did not think highly of American physics, for the most part, and argued for mainly European participants.

After speaking with Jewett, MacInnes decided to write back to Pauli with a friendly dismissal of his concept. Thanking Pauli profusely, he explained that they were seeking an intimate gathering, with up-and-coming researchers, not a large event with venerable professors. Pauli responded by recommending that they speak with Wheeler, who would be more familiar with the newcomers.

As usual, Pauli was right. Wheeler was the perfect choice to bridge the generations. His gentility, deference, and fluency in German had earned him the respect of many of the older European physicists; yet his homespun generosity and quiet humor won him the admiration of young American researchers such as Feynman.

Wheeler graciously accepted the challenge of helping plan the meeting. He relished the idea of distinguished physicists debating subjects close to his heart, such as electron interactions and the role of mesons. Initially, he and Pauli suggested that it be held in Copenhagen under the auspices of Bohr's institute. However, concerned that few Americans would wish to travel to Denmark and hoping to keep costs down, Darrow advocated for a location in the United States. Indeed the total expenses for the meeting would end up being less than $1,000.

After months of planning, the team identified the Ram's Head Inn on Shelter Island as a suitable site. The island would provide an ideal setting for top thinkers to gather in a quiet place and confer.

Just off the northeastern tip of Long Island, it was reasonably close to New York City and southern New England, yet maintained the isolated feel of a rocky outpost in the Atlantic. Dates for the event were chosen so that Oppenheimer, at that point the most famous American physicist and a major draw for the physics community, could attend.

MacInnes worked with Wheeler to put together the list of invited guests. They decided to identify three discussion leaders, each a prominent physicist, to conduct the workshops: Oppenheimer, Kramers, and a brilliant Austrian émigré, Victor "Viki" Weisskopf of MIT.

Weisskopf was another of the many veterans of the Manhattan Project invited to the conference. Like Kramers, he was a protégé of Bohr. Remarkably, he had also worked under Max Born (his PhD advisor), Werner Heisenberg, Erwin Schrödinger, Dirac, and Pauli, making his list of mentors something like a who's who of the quantum world.

Inspired by a suggestion by Pauli, in 1939 Weisskopf had developed an innovative approach to computing the self-energy of the electron (energy due to its interaction with the electromagnetic field it generates), with the ultimate goal of obtaining a finite value. (Like Kramers, but unlike the subsequent Wheeler-Feynman absorber theory, he kept the electromagnetic field.) To obtain a suitable quantum value for the self-energy, he calculated the effects of "vacuum fluctuations."

A vacuum fluctuation occurs when particles spontaneously emerge from the seeming emptiness of space, persist for a short interval, and then sink back into the void, like dolphins rising briefly for air before resubmerging into the sea. For example, an electron and a positron might appear together, enjoy a fleeting glimpse of reality, and quickly annihilate each other. This temporary creation of matter out of sheer nothingness is permitted under Heisenberg's uncertainty principle as long as the duration is sufficiently brief (the larger the mass, the quicker the stay).

Another restriction on such transient entities, known as "virtual particles," is that they must conserve charge. That is one reason positrons emerge along with electrons; their charges cancel each other out. Nature's vacuum is like a credit line that allows considerable borrowing, but with strict limits.

A key feature of the virtual particle sea near an electron (or another charged particle) is that it tends to polarize—that is, align along the positive-to-negative direction in the same way batteries are lined up in a flashlight. This phenomenon is called "vacuum polarization."

Vacuum polarization effectively replaces the point charge of an electron with a cloud of charge. Like the needles of a startled hedgehog, pairs of oppositely charged virtual particles radiate outward from the electron. They do so in such a way that the positive end of each pair is closer to the electron than the negative end. These lined-up charges shield the bare electron's charge such that it effectively becomes a finite clump rather than an infinitesimal point. In essence, it spreads out the electron's mass and charge.

Remarkably, Weisskopf had found that vacuum polarization helped moderate an electron's self-energy—making headway in resolving one of the key conundrums of quantum electrodynamics. The self-energy calculation still blew up to infinity, but less quickly: like a slow termite infestation in a wood house rather than a raging fire; the end was similar, but the process was much slower and potentially manageable. Weisskopf's hypothesis offered promise that further mathematical manipulations might produce a long-sought finite self-energy value.

Along with the three discussion leaders, some two dozen other physicists were invited. Naturally, Feynman was one of Wheeler's picks. Wheeler selected himself too, enabling the two colleagues to enjoy some long-awaited time together for tossing around ideas. Others on the list included Eugene Wigner, John von Neumann, Hans Bethe, Edward Teller, Gregory Breit (who had been one of Wheeler's postdoctoral supervisors), Enrico Fermi, and Julian Schwinger, a brilliant young Harvard theoretician. Like Feynman, Schwinger was originally from New York. He had worked with Oppy and had a stellar reputation. Schwinger and Weisskopf had cemented close links between their respective physics departments, just down the road from each other in Cambridge, Massachusetts.

Not everyone was strictly a theorist. In fact, another invitee, Willis Lamb of Columbia University, was an expert in atomic measurement and would report critical experimental results that would steer the conference's discussion. Rabi, who had been Schwinger's PhD advisor at Columbia and performed both experimental and theoretical research, would similarly present key findings.

THE MIGHTY LAMB AND THE SHIFTING LINES

Wheeler knew Lamb very well and had great respect for his work. In 1939, shortly after Lamb received his PhD from Berkeley, they had published a paper together about how the atoms in the atmosphere affect energetic cosmic rays zooming through it. Cosmic ray electrons, forced by air resistance to yield some of their energy, often decay into other particles in what is called a "cascade." Wheeler and Lamb calculated how atmospheric atoms influenced such particle production. (In 1956, they would find an error in their original calculation, revisit the topic, and publish a sequel article.)

Lamb's best-known research, however, took place in the spring of 1947, while final arrangements for Shelter Island were being made. He and his graduate student Robert Retherford applied a novel way of probing the hydrogen atom and showed that one of Dirac's predictions fell short. In particular, Dirac had calculated that two of the electron states in hydrogen, technically known as $^2S_{1/2}$ and $^2P_{1/2}$, should have exactly the same energy. The Lamb-Retherford experiment uncovered a subtle but important discrepancy in hydrogen's fine structure: a tiny shift in its spectral lines, representing a minute energy difference between the two states in question. The revelation that former atomic models were slightly off offered an opening for theorists to advance the field of quantum electrodynamics.

Through his defense work, conducted at Columbia's Radiation Laboratory, Lamb had become adept at producing concentrated, relatively high-frequency microwaves for the purpose of radar transmissions. Shorter in wavelength than radio waves, these offered more precise means for probing a target, allowing identification of minute deviations in energy levels. After the war, Lamb decided to apply his expert knowledge to tests of atomic properties. In particular, he wished to see if Dirac's model correctly predicted the hydrogen spectrum.

On April 26, following several attempts to apply microwave probes to hydrogen atoms, Lamb and Retherford achieved success.

They found that by applying a signal of approximately 1,000 megacycles (million cycles) per second, they could excite an electron from the $^2P_{1/2}$ to the $^2S_{1/2}$ state. According to quantum principles dating back to Max Planck and Albert Einstein, that particular frequency corresponded to a certain minuscule amount of energy. They thereby proved the existence of a small but distinct energy difference. The discrepancy became known as the "Lamb shift."

As rumors abounded about his discovery, Lamb began to prepare a report for the upcoming Shelter Island conference. He had pushed the study of atomic electrons to much greater precision than other experimenters had ever achieved, resulting in an unprecedented probe of fine distinctions in electron energy levels.

Based on his own calculations, Weisskopf already had a hunch about the energy difference between the two states. He thought that interactions between electrons and the quantum vacuum would cause such a split. Waiting eagerly for experimental proof, however, he had held back on publishing his results. He would later regret his hesitation, believing that it might have cost him a Nobel Prize.

Lamb's presentation would offer the smoking gun showing that, as Weisskopf, Kramers, and others suspected, Dirac's quantum electrodynamics desperately needed revision. Many years later, on Lamb's sixty-fifth birthday, physicist Freeman Dyson would commend him: "Your work on the fine structure led directly to the wave of progress in quantum electrodynamics. . . . Those years, when the Lamb shift was the central theme of physics, were golden years for all the physicists of my generation. You were the first to see that that tiny shift, so elusive and hard to measure, would clarify in a fundamental way our thinking about particles and fields."

CHAPTER FIVE

THE ISLAND AND THE MOUNTAINS: MAPPING THE PARTICLE LANDSCAPE

> "[Feynman] had this wonderful vision of the world as a woven texture of world lines in space and time, with everything moving freely, and the various possible histories all added together at the end to describe what happened."
>
> —Freeman Dyson, *Disturbing the Universe*

There are so many ways to arrive at the same place. Take a dozen brilliant minds, and they will likely figure out two dozen ways to solve the same problem. Working out the problem of quantum electrodynamics took a concentrated effort with many different contributors. The three National Academy of Sciences conferences, held on an island, in the mountains, and in a valley, respectively, revealed several different visions for the particle terrain that theorists would luckily reconcile.

Richard Feynman always followed his own road. His description of quantum electrodynamics would introduce a new lexicon—Feynman diagrams—that would fundamentally change the field. These initially odd-seeming doodles became essential shorthand for modeling particle processes.

The way Feynman added up these doodles, borrowing the sum-over-histories idea from his own thesis work, revolutionized the concept of time in quantum physics. Feynman knew there were various routes to the same end and applied the same flexibility to the subatomic world—only, unlike humans, particles somehow could take multiple paths at the same time.

ISLAND OF DREAMS

On a sunny, mild Memorial Day weekend in 1947, physicists gathered at the American Physical Society headquarters in Midtown Manhattan to prepare for an excursion. Close to Pennsylvania Station, where many would arrive by train, it was a natural meeting place. There, like schoolchildren on a field trip, the physicists were herded into a rickety old bus. Their destination was Greenport, Long Island, to board the ferry to Shelter Island.

After months of planning, John Wheeler looked forward to discussions about the connections between electrons, positrons, and other particles. The conference would offer a great chance to tap the minds of some of the greatest thinkers around, including the unofficial spokesman of the American physics community, Robert Oppenheimer. Maybe he could win some of the attendees over to the "everything is electrons" point of view. At any rate, their reactions would be interesting.

Victor Weisskopf and Julian Schwinger eagerly anticipated Willis Lamb's talk. Informally, they had learned about his exciting results. On the train down from Boston, they had discussed potential applications of Weisskopf's relativistic electron self-energy formula to each of the two states Lamb had targeted. Because Weisskopf's formula blew up at a slower rate, it was more manageable than the nonrelativistic equivalent. Subtracting the two energies would lead to a finite shift of the sort Lamb had found. The specific value just needed to be cranked out to see if it matched.

Where was Feynman? He had either missed the bus or decided not to board it (in a later interview, he wasn't sure why; perhaps he had visited his family). He knew Long Island really well, of course, having grown up in Far Rockaway. So he decided to drive to Greenport.

As the bus meandered through Queens and Nassau, arriving in rural Suffolk County, it passed much traffic going in the other direction. The holiday weekend was coming to an end, and many cars were heading back to New York from various beach locales. Those drivers would have been amazed to learn that many of the designers of the atomic bomb had just passed them in a bus.

Once the physicists arrived in Greenport, however, locals knew something was going on. A police motorcade escorted them through the village, blocking all other traffic as they passed through intersection after intersection. The bus arrived at a hotel where they were to spend the night.

Feynman met the group in a Greenport restaurant to enjoy a free meal, compliments of the owner. After they dined, the owner stood up and applauded the physicists for winning the war. His son had been serving in the Pacific when the Japanese surrendered. He was grateful to the atomic bomb designers that his son had come home safely.

The next day they took the ferry to Shelter Island and arrived at the Ram's Head Inn. The conference was a three-day event, from June 2 to 4. Each morning started with an address by one of the

Physicists at the Shelter Island Conference, 1947. Left to right: Willis Lamb, Abraham Pais, John Wheeler, Richard Feynman, Herman Feshbach, and Julian Schwinger (*Source:* AIP Emilio Segrè Visual Archives).

keynote speakers—successively Oppenheimer, Hans Kramers, and Weisskopf—and then led to more specialized talks. Some who attended the meeting recalled that Oppenheimer was a commanding presence on all three days.

While Lamb's talk, as Schwinger and Weisskopf had anticipated, proved the main experimental highlight, other results presented at the conference also generated much buzz. Working in a different Columbia University lab, I. I. Rabi had enlisted two of his graduate students, John Nafe and Edward Nelson, to carry out a high-precision experiment designed to measure the electron's "magnetic moment," a physical parameter related to how a particle reacts in a magnetic field. Rabi announced at the conference that his team had attained slightly larger results than expected. He also reported anomalous electron magnetic moment measurements taken by his Columbia colleague Polykarp Kusch using different methods.

Gregory Breit, who had been working with Rabi to interpret those findings, gave a talk about his own take on the matter. He had determined that the discrepancy between the original and corrected magnetic moments seemed proportional to an important theoretical parameter called the "fine structure constant," designated by the Greek letter alpha and almost exactly equal to 1/137. The fine structure constant sets the strength of the electromagnetic force by governing the coupling (interaction) between charged particles, such as electrons, and photons. Assuming it is not simply coincidental, such a relationship between the discrepancy and the fine structure constant could demonstrate how virtual photons of the vacuum influence in some manner the magnetic properties of electrons.

With the Lamb shift and the anomalous magnetic moment results pointing to the need for a revised description of the electron, much of the conference's discussion centered on possible ways to fix the problem. Weisskopf offered his suggestions, based on his idea that an electron's neighborhood influences its properties. Schwinger concurred and began to think about ways to rewrite quantum electrodynamics to take into account interactions with the vacuum. He aspired to reproduce those experimental results consistently.

Wedding plans would hinder Schwinger, however. Only a few days after Shelter Island, he'd be walking down the aisle with his bride, Clarice Carrol. A long honeymoon traveling around the United

States would delay his getting down to business. Not until September would he start rewriting quantum electrodynamics in his distinctively systematic and mathematically rigorous manner.

Feynman's talk at Shelter Island took place on the last day of the meeting, when attendees were getting ready to head home. He unveiled what he had begun to call the "spacetime approach" to quantum mechanics: essentially the mathematical methods used in his thesis to describe interactions between electrons, including sum over histories. At that stage, his techniques could not explain phenomena such as the Lamb shift. The concept was not ripe yet for the physics community, so his talk had little impact. Only after the conference would he make some essential changes, apply his notion more successfully, and finally hit his stride—with a theory rivaling Schwinger's but with a distinctive language and a unique view of time.

PARTY TRICKS

When the meeting adjourned, Feynman headed back to Cornell, where he was staying at the Telluride House, a special dormitory for honors students. He was selected as faculty resident, which offered him free food, lodging, and comfortable spaces to do calculations. As he loved hanging out with students, it was the perfect place for him to live.

Hans Bethe returned to Cornell in a more convoluted manner—akin to a less likely path in a Feynman sum over histories. To take care of some consulting responsibilities for General Electric, he stopped for several weeks in Schenectady. On the train ride there from New York, he took out a pencil and paper and began to think of ways to calculate the Lamb shift. Like Feynman, Bethe had a passion for estimates on the fly. As the Hudson Valley slipped past the carriage's windows, more and more symbols and numbers flowed from Bethe's hand. He started with the idea, much discussed at the conference, that interactions with the quantum vacuum influence an electron's mass and energy. Brilliantly, by cancelling out certain terms, he found a way to obtain a finite formulation. When, after his arrival, he plugged in the numbers, he discovered that he had hit the mark: a value close to the measured result of 1,000 megacycles per second.

Strangely enough, Bethe had planned a party that would take place during his absence, to which Feynman was invited. For some reason, when he was diverted to Schenectady, he decided not to cancel or postpone it. Knowing that Feynman would be visiting his house, Bethe decided to call him there. Feynman picked up the phone and heard Bethe excitedly report his results. Given that he had managed to reproduce the measured value, Feynman was intrigued.

Bethe wrote up his Lamb shift results and sent a mimeographed copy to each of the Shelter Island attendees. He also submitted the paper for publication. It wasn't a full-fledged revamp of quantum electrodynamics, as it didn't take relativistic effects into account. Rather, it hinted at a new direction for the field.

Weisskopf was fuming. He'd originally had the idea that the Lamb shift was due to a cloud of virtual particle interactions. He had certainly suggested this at the conference and mentioned it to Bethe. Yet Bethe had presented his results as a single-author work. Weisskopf felt that he should have been listed as coauthor or at least heavily acknowledged. He chided himself for not publishing the idea earlier and getting proper credit.

As physicist Kurt Gottfried, who studied under Weisskopf, recalled, "Viki . . . felt that Bethe received too much recognition, but he also recognized that he really had not completed such a calculation when Bethe submitted his work for publication. He did for quite some time remain unhappy about this topic."

After Bethe returned to Cornell, he gave a seminar about his results. Feynman, who was in the audience, saw that while consistent with the Lamb shift, Bethe's calculations still had a haphazard quality. They involved several cancellations of infinite terms that Bethe couldn't adequately justify. He needed to get rid of the infinite values to obtain a realistic, finite answer. Yet he couldn't explain why certain ones happened to cancel each other out to produce, in something like a magic trick, the expected value of 1,000 megacycles per second.

One plus infinity equals infinity. Two plus infinity equals infinity. In fact, any number plus infinity equals infinity. Therefore, infinity minus infinity equals any number at all. How could Bethe justify that his own subtraction of infinities happened to hit the bull's-eye?

Bethe had done the calculation nonrelativistically. Feynman told Bethe at the end of the seminar that he thought he could bring it in line with special relativity. The next day, he stopped by Bethe's office with his preliminary attempts. They weren't quite right yet, but he was persistent. He applied his genius during the ensuing months to a completely novel approach to quantum electrodynamics, one that matched the Lamb shift and other experimental results.

SQUIGGLES, LINES, AND LOOPS

Applying the toolbox of techniques he had picked up at Princeton, Feynman set out to understand how electrons interact with the quantum vacuum in a manner that alters their observed mass and charge. He needed to take several steps, however, to bring his methods in line with what was considered normal. First, he needed to banish his notion, dating back to MIT, that electrons don't have self-interactions. In fact they do, as evidenced by the possibility that they can emit a virtual photon and gobble it up again. That was the idea behind Bethe's calculation, and Feynman knew that he needed to embrace it. He dropped action at a distance and reverted to the more standard concept of electrons interacting via photons.

Second, he needed to abandon the idea of backward-in-time signals and reverse causation. Bethe hadn't included such a possibility, so for consistency he didn't want to either. He wasn't sure if he really believed in them anyway, even if the absorber theory required them. Consequently he replaced the fifty-fifty mixture of backward and forward with only temporally forward signals. In other words, photons traveled exclusively into the future, and cause always preceded effect.

Feynman retained two important aspects of his earlier work, however. Although he dropped backward-in-time signals, he kept backward-in-time electrons as stand-ins for positrons, as Wheeler had suggested. The only alternative he had seen for dealing with positrons was Paul Dirac's hole theory, which was notoriously hard to handle in detailed calculations. Wheeler's time-reversed electrons were crazy to think about but computationally much easier to consider. Feynman simply ignored the "all electrons are the same one" aspect of Wheeler's hypothesis and kept the most useful part: its simple way of representing positrons.

Even more significantly, Feynman continued to apply the sum-over-histories techniques that he had mastered, recasting quantum electrodynamics within the framework of his spacetime approach. His way of tallying every single conceivable quantum path according to a quantity called its "amplitude," obtained by multiplying transformation functions and squaring those amplitudes to get probabilities, proved a clever and powerful way of describing how quantum fields interact in particle physics.

Wheeler had drummed into Feynman's head the importance of drawing diagrams to represent physical phenomena. Feynman loved art and found visual representations extremely useful. So he came up with his own visual shorthand for depicting the possibilities for particle interactions. Representing space along one coordinate axis and time along the other, he'd sketch the essential features of how the particles behaved. As a bonus, spacetime diagrams were ideal for including the effects of special relativity. These pictorial representations—subsequently developed further through discussions with Freeman Dyson about graphs in mathematics—would come to be known as "Feynman diagrams."

Feynman's early diagrams were relatively primitive and had an ad hoc quality. Over time, in consultation with Dyson, he would establish more consistent rules. Dyson's contributions to the diagrammatic method would prove so vital that many early references gave him joint credit.

In the final version of Feynman diagrams, electrons (and other material particles) are depicted as directed line segments pointing forward in time. Typically, arrows indicate their directions. Positrons are shown as directed line segments too, but oriented backward in time. Squiggles represent photons, as they are emitted or absorbed by charged particles or when they veer off into space. A loop shows a virtual electron and a virtual positron (or another oppositely charged particle-antiparticle pair) emerging from the vacuum and then returning to it again by annihilating each other. Another way of looking at that loop process is that the virtual electron keeps circling through time—backward and forward, again and again. In that manner, the closed cycle acts as a kind of quantum ouroboros (a serpent eating its tail).

Therefore, for an electron giving off a photon that is picked up by another electron, a diagram would show directed line segments for each incoming electron, a squiggle connecting their paths, and differently angled line segments for the outgoing electrons. The change in the electrons' directions would result from a photon exchange. The arrows for each electron would point forward in time; otherwise the diagram would depict positrons.

To take into account sum over histories, Feynman would draw a different diagram for each conceivable process. Of course, some were less likely than others. Still, the sum had to include every possibility, weighted according to its likelihood. Therefore, a typical particle event, such as scattering, would involve a set of diagrams, not just one. He also tallied the possibilities for different quantum spin numbers, another essential factor.

To represent how vacuum effects produce the Lamb shift, Feynman drew a directed line segment to represent the bare electron. He called it the "direct path." Then he linked a squiggle to the segment, showing what would happen if the electron emitted and then reabsorbed a virtual photon. He referred to the addition of the photon as a "correction," which took into account a certain type of quantum fluctuation. That was one possible history.

Another diagram, which Feynman would consider later, involved a loop connected to a line segment by a squiggle. It depicted an electron emitting a virtual photon, in turn producing a virtual particle-antiparticle pair. Adding these two diagrams obtains the first-order (most basic) correction to the electron's energy, approximating the spectral shift detected by Lamb and Robert Retherford. (Getting the second-order correction involves diagrams with two loops, and so forth.)

In his zeal to get the right values, however, Feynman took a lot of liberties with his math. While he had developed a clear diagrammatic method, he didn't check its mathematical rigor. Nonetheless, it did seem to work—with adjustments in a few parameters. It was like jury-rigging a radar system by connecting a bunch of wires and being satisfied when blips started appearing on the screen, without considering whether the makeshift contraption satisfied the electrical code.

"You know how I work so most of it is just a good guess," Feynman would later write to physicist Ted Welton of the University of Pennsylvania, an old friend from his MIT days. "All the mathematical proofs were later discoveries that I don't thoroughly understand but the physical ideas I think are very simple."

Feynman cared mainly about whether the end result matched the experimental prediction. Mathematicians and philosophers might waste their breath arguing about the rhyme and reason, but not he. As long as his diagrams got the correct answers in a predictable way, he was happy. As scientific historian David Kaiser has remarked, "Feynman believed fervently that the diagrams were more primary and more important than any derivation they might be given. In fact, Feynman continued to avoid the question of derivation in his articles, lecture courses, and correspondence."

Luckily Dyson, who joined Cornell as Bethe's graduate student, would help develop and promote Feynman diagrams while linking them more closely to other interpretations of quantum electrodynamics. His unification of the various methods would prove critical for their success.

AN ENGLISHMAN'S ODYSSEY IN ITHACA

As a boy in the south of England, Freeman Dyson spent his holidays at a seaside cottage. His father was a classical musician whose most famous work is "The Canterbury Pilgrims," and his mother was a Chaucer enthusiast, so his love of books and quiet contemplation did not surprise them. Nevertheless, they were baffled when one Christmas vacation he brought along a book on differential equations and read for an average of fourteen hours every day. He worked out all the problems, hoping to fulfill the text's promise of mastering Albert Einstein's theory. It was his favorite Christmas ever!

Through a Commonwealth Fellowship, the graduate of Winchester College (an English private school) and Cambridge ended up spending a year in the Cornell department headed by Bethe and often serenaded by a certain bongo drummer named Dick Feynman. The recluse and the rascal got along famously; Dyson and Feynman quickly became close friends and confidantes. Most importantly Dyson would help bring Feynman's work to the physics community's attention by promoting his diagrammatic approach, while aspiring to firm up its mathematical basis and connecting it with the related work of Schwinger.

When Dyson arrived as a graduate student in September 1947, Cornell's informality startled him. Everyone called Bethe "Hans"—unheard of at a British university of the day. Dyson was also taken aback by the famous physicist's casual style, particularly the muddy shoes he wore at their first meeting. If a Cambridge don were seen in such footwear, people would assume that he had fallen into a ditch. In the United States, that was just his normal gear.

If Hans was unconventional, Dick was insane, but demure Freeman loved it. He had seen nothing in his life like the young genius from Far Rockaway, with his drumming at 2:00 a.m. and his wild sense of humor. Feynman flaunted convention in a way Dyson had never even imagined, especially from a professor.

Dyson found Feynman's methods similarly bizarre at first. Yet, strangely enough, they worked like a charm. They not only reproduced Bethe's results but also predicted things that Bethe couldn't calculate. As in Edith Nesbit's *The Magic City*, another book Dyson had loved as a child, marvelous things seemed to emerge from thin air in a universe with crazy rules.

The madman ordains that bongos must be played at night. Everything that can happen does. Electrons love to do loop-the-loops in time or couple with squiggles from the void. Reality must encompass every possibility. Add that all up, and abracadabra, hocus-pocus, you can predict hydrogen spectral lines.

Dyson recalled the first time he encountered Feynman's peculiar approach to quantum physics: "Hearing about the sum over histories directly from Feynman at Cornell in 1947, I was amazed and bewildered. Amazed because the picture was physically right. Bewildered because it was mathematical nonsense."

It would be quite some time before Feynman took the trouble to lay out the mathematics behind his methods. In the meantime, he would calculate enigmatically with diagrams, like a conjurer hoarding his secrets. He relished the look of wonder on people's faces when he got the right answer, without even seeming to perform a calculation. Dyson, in contrast, while bringing the diagrammatic method to a wider audience (a variation was even called "Dyson diagrams" for a time), pushed much harder for a way to render them more mathematically sound.

Mathematician Marc Kac once compared Bethe and Dyson with Feynman, stating, "In science . . . there are two types of geniuses: the 'ordinary' and the 'magicians.'" While Bethe and Dyson were truly brilliant at calculations, they took clear, straightforward steps. Feynman, in contrast, was no "ordinary genius" but rather, in his seeming ability to pull results out of thin air, "a magician of the highest caliber."

MARATHON ON THE SUMMIT

The sequel to the Shelter Island conference, held at the Pocono Manor Inn in the Pocono Mountains of Pennsylvania from March 30 to April 2, 1948, would prove just as important—albeit more for its introduction of new theoretical methods than for the presentation of novel experimental results. Schwinger would dominate the meeting, presenting his meticulous, mathematically rigorous version of quantum electrodynamics, which offered considerable progress toward cancelling infinite terms and producing finite answers. With far less fanfare, Feynman unveiled his diagrammatic, sum-over-histories method as an alternative approach. Several new faces at the conference hadn't been at the previous meeting, notably Niels Bohr (and his son Aage), Eugene Wigner, and Dirac.

Nicknamed "The Grand Lady of the Mountains," the historic inn proved a comfortable setting for the meeting. Oppy once again ran the show. For American theoretical physics, he was still the unchallenged MC.

The attendees settled in for long days of talks, particularly the day when Schwinger presented. Speaking for around eight hours, he put equation after equation on the board, methodically detailing a comprehensive approach to quantum electrodynamics that matched new experimental findings. He drew from Kramers's concept of renormalization—redefining mass and charge to eliminate infinite terms—without adopting Kramers's sheer dismissal of Dirac. Instead, Schwinger generalized Dirac's relativistic quantum concept of the electron to include its electromagnetic interactions in such a way that it consistently gave reasonable, finite answers. The tweaks to mass and charge offered just the trick for cancelling the very terms that were causing problems, vanquishing the dark clouds of divergence and leaving clear skies ahead.

Obtaining a finite result from an infinite tally might at first seem impossible. However, sometimes the answer one gets is a matter of how one arranges items into groups. Alternative ways of grouping and counting might reveal how a seemingly infinite sum is really finite.

Let's consider, for example, a whimsical situation in which Peter Pan and Wendy, living together forever in Neverland, exchange gifts every day. Suppose he gives her three gold trinkets every morning, and she gives him six silver trinkets every evening. As six silver trinkets are worth exactly three gold trinkets in Neverland currency, it is an even trade. By any reasonable reckoning, Peter's net gain or loss of funds each day adds up to zero.

Suppose that one day Peter loses his own reflection, putting him in a bad mood. He decides to take stock of his life. He tallies the number of trinkets he'll end up giving Wendy throughout eternity and realizes that it will be an infinite sum. After he confronts her, she does the same and realizes that she will be giving him an infinite number of trinkets too. They are extremely cross with each other, until luckily Schwinger Bell—er, Tinker Bell—arrives and shows them how to group the terms by days to obtain a finite answer.

Each day, Tinker Bell points out, their net exchange is worth zero. Therefore, by adding these zeroes together day by day, the total result is finite: simply zero. The infinities were just the result of a poor counting method. Satisfied by the explanation, Peter and Wendy live happily ever after in egalitarian bliss.

We have no record of whether during Schwinger's marathon talk anyone grew sleepy enough to drift off into whimsical daydreams. The mathematics, though solidly and meticulously presented, must have seemed magical enough. His deft techniques for calculating known experimental values such as the Lamb shift and anomalous magnetic moment of the electron were truly a marvel. Even presenting for so long was a feat. Wheeler, who was taking careful notes for the conference, compiled forty pages detailing what Schwinger discussed, far more than for any other speaker.

Schwinger's accomplishment, based on months of intense effort, was truly impressive. Starting in September 1947, he had become a one-man theoretical powerhouse. With his wedding and honeymoon over, he had launched full steam into a systematic exploration of ways to workably formulate quantum electrodynamics. Weisskopf's findings on the role of vacuum polarization and Kramers's concept of distilling an electron's bare mass from its experimental mass via renormalization proved key to his calculations. Schwinger also made heavy use of the mathematical fields known as "group theory" and "gauge theory," explored by Wigner, mathematician Hermann Weyl, and others. Schwinger's full mastery of the subject would shine through in his talk.

The unusual thing about Schwinger's work, starting from around that time, was that he wanted to do everything himself. There were no sous-chefs, only the head chef. Ironically, throughout his career he would take on tons of graduate students—far more, for example, than Feynman. Yet he would essentially press his students to work on their own.

Other physicists would joke about how Schwinger aspired to dominate the field using his own methodology. The *Journal of Jocular Physics*, a series of humorous volumes celebrating various birthdays of Bohr, included a satirical guide to successful publications in physics. The mock template began with the phrase, "According to Schwinger . . . " and advised researchers to fill in the blanks afterward. Schwinger's carefully crafted talk at the Pocono conference hinted at such attempted domination to come.

DICK'S DEBUT DEBACLE

By the end of Schwinger's epic presentation, attendees were impressed but exhausted. Unfortunately, Feynman was scheduled to speak next. Although normally a highly engaging speaker, in this case he was ill prepared to explain his exotic new ideas. Instead of starting from basics—the least action principle, integrating over different paths, and so forth—he acted as if everyone knew the theory and was primed for an example. Therefore, those at his talk felt like they had missed the first half of an advanced physics course and were attending a problem-solving review session. They were not only tired but totally lost.

The main example Feynman chose consisted of two electrons interacting by means of a virtual photon. After he rapidly tried to explain how to calculate their behavior using his diagramming method, the bleary-eyed audience had practically no idea what he was doing. He rushed through the idea of sum over histories and the concept that positrons are electrons traveling backward in time. Wheeler, intently taking notes, could at least follow those parts. The rest grew even more confused.

Most of the audience members knew of Feynman's brilliant contributions to the Manhattan Project. Some were also aware of his tragic loss of Arline and cognizant of the trauma inflicted by the war and its tough decisions. Given that Feynman had undoubtedly been depressed, some must have wondered if he had completely lost it. Had the genius gone bonkers?

After Feynman's talk, Bohr led a discussion session. In his own characteristically subdued way, he offered a stinging critique. Not fathoming Feynman's shorthand, he pointed out that Heisenberg's uncertainty principle made it impossible to draw the exact trajectories of electrons through space over time. How could Feynman purport to sketch electrons' actual paths? Bohr also dismissed the notion that electrons could travel back in time, completely contrary to basic physical principles. Edward Teller was aghast that Feynman's treatment didn't appear to comply with the Pauli principle preventing two electrons from ever residing in exactly the same state. Wheeler took careful notes but, as far as history records, did not defend Feynman. Wheeler always treated Bohr, his dear mentor and role model, with the utmost deference and respect.

Dirac, best positioned, along with Wheeler, to examine the sum-over-histories aspect of Feynman's talk, as some of it was based on his own work, raised an important question. In summing the weighted amplitudes of the various paths, did Feynman check that the total probability for all the options added up to 100 percent? Embarrassingly, Feynman replied that he hadn't. This was a significant blunder, as the tally could be either missing or overcounting certain modes of interaction. It was like saying that there are 50-50 odds of a coin being under each of three cups, adding up to a 150 percent chance of it being under any one of them. It just doesn't compute. Feynman needed to check the total and adjust the odds if necessary (which it was).

Another critique of Feynman's presentation held that he had failed to take into account vacuum polarization—the great innovation of

Weisskopf that had become an essential part of Schwinger's calculations. Feynman's notation represented it with a one-loop diagram. For brevity, he decided to present something more basic, just two electrons exchanging a photon, but this did not fulfill the audience's desire for an explanation of the key experimental findings—the Lamb shift and anomalous magnetic moment. Schwinger clearly could do it; Feynman seemed like he was just drawing pointless doodles.

Bethe sympathized with Feynman, who left the meeting feeling awful. Feynman resolved to write up his ideas as an article and explain his techniques in a much more thorough fashion. As Bethe recalled,

> Feynman had a completely new way of looking at things—which I knew but most of the other people found strange. And especially Niels Bohr, who, after all, was the leader of us all. Niels Bohr couldn't understand it, wouldn't believe it, gave some very sharp arguments against it, and treated Feynman rather badly. And Feynman, of course, was very much disappointed because he had what he considered a beautiful theory. And here was the greatest of all quantum physicists, who wouldn't believe him. So when he came home, I had to console him. I was at the meeting; I heard the presentations, as well as Bohr's reaction. Unfortunately, Feynman likes to present his work—or did, at that time, like to present it—as paradoxically as possible. And this was just impossible for somebody like Bohr to understand.

WILD ROAD TRIP

Following his months-long burst of creative genius, which had boosted his spirits, Feynman's sense of self-worth plunged again. By being inadequately prepared for the questions raised by his talk, he had humiliated himself in front of Bohr, Dirac, Oppy, and so forth—the very royalty of quantum physics. His life had become like the Coney Island Cyclone, with exalted highs and stomach-churning dips. In his darkest moments he felt that his calculations were pointless anyway, given that they wouldn't stop atomic bombs from proliferating and wiping everyone out. In fact precisely his calculations at Los Alamos had helped open that Pandora's box. It was senseless looking back, but, still, he and the world had really goofed.

As usual, when he felt so incredibly stupid, he channeled his creative energy into his teaching. He felt much more confident explaining things to students than to glaring physics icons. Furthermore, leading young minds to discovery was a far better use of his time than sulking, which he never liked to do. When he had trouble sleeping, banging on the bongos for a while also relieved his stress.

Another pleasant distraction, when he managed to slip off campus, was scoping out pretty young women from any walk of life—it didn't matter what they thought about or knew about physics. Youthful and handsome, he also courted college students, who were often surprised to learn he was a professor.

He had kept in touch with a woman from New Mexico, a secretary he'd met during his Los Alamos days. She had begun, in their correspondence, to refer to another guy she seemed to like. With hopes of visiting her, heading off the other guy's overtures, and sparking a flame before it was too late, Feynman decided to take a road trip to Albuquerque. Besides, he really missed the freedom of the Wild West and wanted to reexperience it. The summer of 1948, after classes ended, would be a perfect time.

Also, as it turned out, Dyson needed a ride westward. He had signed up for a summer seminar in Ann Arbor, Michigan, offered by Schwinger as a kind of crash course in his methods of quantum electrodynamics. Bethe strongly encouraged Dyson to learn such techniques.

Feynman offered Dyson a lift to Albuquerque. From Dyson's point of view, it was far from the most convenient option. First, the class wouldn't start until August, and Feynman was leaving in June. Moreover, Ann Arbor and Albuquerque were close alphabetically, not physically; on the contrary, they were thousands of miles apart. It was certainly not the kind of efficient spacetime trajectory an elementary particle would take—at least a sane, sober one.

Well, Fermat's principle of least time be damned: Dyson wanted to explore the American West, even in a completely convoluted manner. He decided to journey to Albuquerque with Feynman—four days of driving—and then take a long-distance bus back to Ann Arbor when it was time for his course. The Commonwealth Fellowship nicely included summer travel money, so he might as well use it.

In the days before interstate freeways, the best road from the Midwest to the Southwest was Route 66. The winding highway connected countless towns where it typically served as Main Street. Service stations, cheap hotels, smoky bars, and seedy diners broke the monotony.

The long journey cemented the two men's friendship. As they passed endless fields of grain and the gentle rolling vistas of the Ozark Mountains, they talked about their backgrounds, passions, and views of physics. Every once in a while, they'd pick up someone thumbing a ride and drop him off sometime later. Not even the grimiest hitchhiking vagabond rattled Feynman. Never knowing what to expect but enjoying himself immensely, Dyson trusted Feynman's judgment.

As they approached Oklahoma City, suddenly the sky opened up with a severe downpour. When the road became too flooded, they turned around, drove back to the town of Vinita, and desperately looked for a place to get out of the rain. As other stranded drivers had the same idea, there weren't many good choices. Feynman decided to stop at a fleabag hotel offering shared accommodation for fifty cents a night. There was no door to the room, only a hanging sheet. The behavior of some of the young ladies roaming the corridors led Dyson to realize that the place doubled as a kind of brothel. He was too nervous to venture outside the room, even to use the bathroom down the hall. With nowhere else to go, they decided to cloister themselves in their enclave and make the best of it.

With the rain pounding down outside, the two physicists poured out their hearts and revealed to each other their inner perspectives. After talking about life at Los Alamos, Feynman confessed his feeling that it was only a matter of time before nuclear weapons would destroy the world. Dyson was surprised by Feynman's dispassionate expression of this apocalyptic view. He was a coolheaded Cassandra, convinced of coming disaster but calmly carrying on. Feynman was the sane one, Dyson realized; the rest of the world was crazy.

Their discussion then turned to quantum electrodynamics. Feynman expressed his hope to extend his methods to all known forces, finally achieving the unity that had eluded Einstein. Given that sum over histories works for electromagnetism, it should apply to nuclear interactions and gravitation as well. Feynman articulated a grand vision in which a medley of diagrams expressing all conceivable histories could map out everything in nature.

Dyson countered that it would be better to focus on electrodynamics first and bring the various quantum approaches in line with each other: Feynman's, Schwinger's, and a third method developed during the war by Japanese physicist Sin-Itiro Tomonaga and conveyed to Bethe and Oppy shortly after the Pocono conference.

Tomonaga's concept, somewhat like Schwinger's only more simply expressed, had been published in Japanese journals in 1943 but became available in translation only in 1948. Astonished by Tomonaga's insights, given his isolation from the Western physics community during the war, Bethe, Oppy, and others working in the field were also stunned that he made his breakthrough well before the discovery of the Lamb shift.

When the rain let up the next day, Feynman and Dyson drove onward to Texas, spent one more night together, and then headed to Albuquerque. Approaching the city they hit a speed trap, got pulled over, and received an astronomical ticket. Sent to the courthouse, Feynman luckily managed to convince the justice on duty to lower the penalty.

After paying half the speeding fine, Dyson bid adieu and hopped on a Greyhound bus to Santa Fe. Feynman stayed in Albuquerque, only to find that the woman he had been corresponding with was unavailable. Nevertheless, he remained in the city for some time, trying his luck with the ladies at various bars and honky-tonks up and down Route 66.

THE OMNIBUS APPROACH

From Santa Fe, Dyson hopped on bus after bus, gradually heading east to Ann Arbor. He could pace himself, as he had plenty of time until Schwinger's course began. As wheels spun on prairie roads and the sun sank beneath the flat terrain, he struck up memorable conversations with his fellow passengers about their passions and experiences.

The five-week class proved eye-opening. During each lecture Schwinger's chalk skated marvelously across the board—left to right, top to bottom, again and again—tracing brilliant calculations. Each step followed from the previous one with ironclad mathematical logic. His expositions fell short only in that they couldn't easily be pictured and thereby seemed more like math than physics. Luckily Dyson got on well with Schwinger and had many opportunities to ask him privately about his underlying thought processes.

Dyson worked through his notes carefully, until he had mastered Schwinger's techniques as well as he knew Feynman's. He saw a key difference between the two ideas. Schwinger's was like a seamless tapestry so perfectly woven that it conveyed little about its motivation and purpose. If a stitch came loose and Schwinger weren't available, the whole thing would come unraveled, as no one else would dare try to fix it. Feynman's, in contrast, was akin to a primitive painting layered sensually with thick brushstrokes, smeared and snarled in places but far more inviting. Sure it had flaws, but it was much more accessible. Others might have a go at using his techniques to create similar works. Might there be an artful fusion of the two approaches, melding thought and feeling into a powerful, vibrant theory?

Following his stay in Ann Arbor, Dyson decided to travel around the country some more on his own by bus. He took a boomerang route, first heading west to San Francisco via the majestic Rockies and then shooting back to the East Coast. The long trip offered him plenty of time to think deeply about ways to reconcile the different methods.

On the last leg of his voyage—several days and nights zooming eastward along the highway—a unified vision of quantum electrodynamics crystalized in his mind. To his great delight, he realized that one could write Schwinger's formalism in such a way that it naturally split into a series of Feynman diagrams. In other words, the terms could be organized to represented a sum over histories. He had "Feynmanized" Schwinger! Given that Tomonaga's work bore much in common with Schwinger's, he found that he could bring it in line with Feynman's conception as well. Aboard that clunky Greyhound—while the passengers packed in the seats around him read mystery novels or completed crossword puzzles—Dyson had solved a much deeper riddle: how to unify particle physics.

Finally, after his exciting summer of travel and revelation, Dyson arrived in Princeton to begin a yearlong stay at the Institute for Advanced Study (IAS). Oppy, its new director, had appointed him a visiting member from fall 1948 to spring 1949, giving him a good

Freeman Dyson at the Institute for Advanced Study in Princeton, New Jersey (*Source:* AIP Emilio Segrè Visual Archives, Physics Today Collection).

opportunity to complete the fine details of his unification of the different varieties of quantum electrodynamics. He'd already thought of a title for the article he planned to write: "The Radiation Theories of Tomonaga, Schwinger and Feynman." He would quickly complete the comprehensive paper, submitting it to *Physical Review* in October 1948. Published early the next year, it would prove highly influential.

At the IAS, Dyson would have the chance to discuss his ideas with other brilliant young minds, notably French mathematical physicist Cécile Morette, a woman fiercely independent in her perspective. Her support for the sum-over-histories method would help affirm its validity.

INTEGRAL TRUTHS

Like Feynman, Cécile Morette knew tragedy at a young age. Only in her case, she faced the death of loved ones earlier in life. During the Nazi occupation of her country, she lived with her family in Caen, a city in the region of Normandy. Wanting to visit Paris to "have adventures," she decided to enroll in advanced courses in mathematics at the city's university. Otherwise she would have been prohibited from traveling there, due to the restrictions imposed by the occupying regime. While she was in Paris, the D-day invasion of Normandy took place. As part of that operation, the Allies bombed Caen, aiming to clear the area of German tanks and troops. Tragically, thousands of French civilians perished. Morette was taking an exam when an errant bomb struck her family home and killed her mother, sister, and grandmother. She was only twenty-one at the time.

Suddenly forced to deal with life on her own, Morette completed her PhD, writing a thesis in mathematical physics on the scattering of mesons. After she had traveled to institutes in Dublin and Copenhagen for research positions, Oppenheimer invited her to become a visiting member of the Institute for Advanced Study. She would work there from 1948 to 1950.

During her first year at the institute, Dyson also arrived for a visiting position. (He would later become a permanent member.) He remained incredibly excited about Feynman diagrams and the sum-over-histories method. Yet its shaky mathematical foundations were all too obvious. He discussed his concerns with Morette, who thought otherwise. She was optimistic about proving the technique valid. As Dyson recalled, "After moving to Princeton in 1948, I talked frequently with Cécile Morette (later DeWitt), who was convinced she could make the sum over histories mathematically rigorous. I argued with her strenuously and kept an open mind."

Indeed Morette would use her mathematical savvy to help place the method, formally known as "functional integrals" or "path integrals," on more solid mathematical footing. Mathematical physicists familiar with her work would apply the technique to many types of problems hard to solve otherwise. As Barry Simon, who wrote a book on the topic, related, "It seemed that path integrals were an extremely powerful tool used as a kind of secret weapon by a small group of mathematical physicists."

In October 1948, Dyson and Morette took a ten-hour train ride from Princeton to Ithaca to spend a weekend with Feynman. He met them at the station on Friday evening, took them back to his place, and regaled them until 1:00 a.m., pounding various rhythms on

Native American drums that he had picked up in New Mexico. The next day, Feynman amused and astounded Morette with a crystal-clear explanation of his theories and amazed Dyson by solving two seemingly intractable problems. As Dyson recalled, "That afternoon, Feynman produced more brilliant ideas per square minute than I have ever seen anywhere before or since."

Soon the weekend drew to a close, and Feynman wished Dyson and Morette a fond farewell. They returned to Princeton gushing with even greater enthusiasm about the power of Feynman's methods.

Feynman published five key articles around that period. The first, "Spacetime Approach to Non-relativistic Quantum Mechanics," introduced his sum-over-histories approach far more palatably and coherently than had the Pocono talk. Another, with Wheeler, furthered their explanation of their joint absorber theory. Yet another piece put forth the idea that positrons are electrons moving backward in time and showed that theory's advantages over Dirac's concept of holes. Two more papers detailed how Feynman introduced relativistic effects into his conception.

At Oldstone, the third quantum field theory conference sponsored by the National Academy of Sciences, held near Peekskill, New York, from April 11 to 14, 1949, Dyson debuted his unified scheme. His talk, as well as a far more coherent explanation by Feynman, demonstrated how Feynman diagrams offered an incredibly practical way of handling particle interactions. Thanks to the talks, and all the important articles by Feynman and Dyson published around that time, the diagrams' elegant simplicity rapidly caught on. Soon they were an indispensable part of any particle theory paper. The concept of sum over histories, opposed (or at least ignored) by Schwinger, would take much longer to achieve wider acceptance among the mainstream physics community.

Once accepted, however, sum over histories would prove a profoundly important step forward for understanding quantum mechanisms. It helped place quantum theory within the context of centuries-old thinking about why objects travel along the paths they do. They do so for the same reason people buy maps (and later GPS devices): to make their routes more efficient. The only difference between classical and quantum is that in the latter the less efficient routes exert a ghostly influence too—like the alternative suggestions in a mapping program. Computations of the results of quantum processes such as scattering need to weigh all the disparate paths, not just the one with the greatest amplitude (which would be chosen in a purely classical system). As Dyson related, "The sum over histories finally gave us an intuitive picture of quantum processes, clear and simple enough for beginning students to grasp. It can be understood also by people who are ignorant of calculus."

Shortly after Oldstone, Dyson's stay at the IAS ended. He returned to England, where he joined the University of Birmingham as a research fellow. He would remain in that position until the spring of 1951.

Meanwhile, in the fall of 1949, a bright young American researcher, Bryce Seligman DeWitt, joined the IAS. Soon he became interested in the adventurous and independent Cécile. The last thing on her mind, however, was getting bogged down in a relationship. She fully intended to return to France after her visiting position was over. Love had other plans.

QUANTUM GRAVIDYNAMICS

Bryce DeWitt had come to the IAS from Harvard, where he was completing his PhD under Schwinger. There, he had been trying unsuccessfully to bring gravitation under the umbrella of quantum theory using methods akin to quantum electrodynamics. He called his concept "quantum gravidynamics."

Schwinger had brought DeWitt's attention to prior attempts to quantize gravitation by other physicists such as Leon Rosenfeld. In particular, Rosenfeld had tried to calculate the gravitational self-energy of a photon and found it to be infinite. Schwinger approved DeWitt's thesis topic with the hope that his newfound renormalization techniques would banish such infinities. After their initial meeting, though, and a few more brief discussions, Schwinger did not spend much time with DeWitt, preferring to focus on his own projects. As DeWitt recalled his experience with Schwinger,

> Well, I probably saw him a total of 20 minutes altogether. I tried to put gravitational fields in quantum electrodynamics, got bogged

> down in quite a lot of difficulties, and went to see him. He said try removing the electron-positron field and just switching on gravitation and electromagnetism. He knew the literature and there had been a paper around 1930 by Leon Rosenfeld showing that the gravitational self-energy of the photon was infinite. And he thought that was nonsense—it cannot be infinite and maintain gauge invariance, to correspond to a renormalization of charge. Anyway, so then he let me go ahead on that. And I think I got good recommendations from him later because I left him alone.

As DeWitt had discovered, Schwinger's methods wouldn't work for gravitation. One couldn't treat gravitation as just another field and expect to churn out finite results. Despite his and many others' attempts, a quantum field theory of gravitation could not be renormalized using the mathematical techniques that Schwinger had pioneered. The infinite terms persisted like an indelible stain.

Despite his setbacks at Harvard, DeWitt's experience propelled him on a lifelong quest to bring gravitation into the quantum fold. Quantum principles couldn't just pertain to some of the forces, he surmised; they must apply to all. At that early stage he was "pretty green," as he put it, and thought the reconciliation would be much easier than it turned out to be. In fact, despite many decades of effort, at the time of this book's writing, it still hasn't been done. "I was just a student and I didn't know the world around me very much," DeWitt recalled.

Once at the IAS, the charming young physicist soon grew close to Morette. After a long, fun day canoeing together, he decided to propose to her. She turned him down at first, as she fully intended to return to France. However, soon she thought of a way the marriage could work and still fulfill her dreams of a homecoming. She was willing to marry him and reside in the United States if they could find a way to spend part of the year in France. Surprised and delighted by her reconsideration, DeWitt agreed to her terms. As a result of that pact, soon after they wed in 1951 she would establish in the French Alps the Les Houches Summer School, a leading incubator for important ideas in theoretical physics.

THE TRAITOR AT THE TABLE

On September 23, 1949, a startling announcement by President Harry Truman changed the world forever. The Soviets had tested an atomic bomb, inaugurating an arms race between the two superpowers. Given the Herculean efforts needed to complete the Manhattan Project and how quickly the Soviets had exploded their own device, many pundits wondered if they somehow obtained access to highly classified information. Several months thereafter, the unmasking of a spy would at least partly answer the question.

Klaus Fuchs, one of Feynman's closest friends during their Los Alamos days (who often lent Feynman his car), had been secretly working for the Soviets since the Manhattan Project began. He had sent them detailed blueprints for the atomic bombs constucted and deployed, as well as sketches of some of the concepts under development, such as Teller's ideas for a superbomb based on hydrogen fusion. Having those plans, particularly for the finished bombs, likely gave the Soviets a jump start of at least a couple years. Eventually they would have built a bomb, but without Fuchs's clandestine help, clearly there would have been a greater time lag.

Fuchs had completely fooled Bethe, Oppy, Wheeler, and everyone else connected with the project. Teller, who greatly distrusted the Soviets, was particularly upset, fueling his determination to press for even more powerful weapons. Fuchs had even been present when they had discussed the possibility of security breaches, and no one had suspected him. Feynman, with his deliberate attempts to circumvent protocol and break into safes, had acted far more suspiciously. He reportedly once jested that if anyone were a spy, it would be him.

After the war, Wheeler had joined an advisory group called the Nuclear Reactor Safeguard Committee, which offered recommendations about the development of commercial nuclear power. For example, it recommended the use of domes as a safety measure for power plants. Teller served as the group's chairman. Feynman also belonged but attended only the first meeting.

Before his unmasking as a spy, Fuchs had attended one of the committee's sessions, held jointly with British physicists in the Harwell laboratory near Oxford. Fuchs was a UK resident, having lived there before and after the war. As a top atomic scientist who could provide valuable advice about nuclear safety, he was welcomed at the joint meeting without hesitation.

Wheeler vividly recalled how he raised at the meeting the possible threat of power plant saboteurs. Someone with inside knowledge could disrupt safety mechanisms and trigger what we now call a nuclear meltdown. In his view, for a system with many safeguards, sabotage was a more formidable risk than accidental failure. Such a saboteur could be anyone, most likely a political zealot with considerable expertise, quietly determined to generate maximum mayhem. As Wheeler described this profile of betrayal, Fuchs sat stone-faced on the other side of the table.

After the Soviet bomb test, Scotland Yard launched an extensive investigation into possible security breaches. The trail led to Fuchs, who was arrested on February 2, 1950. He confessed to passing atomic secrets. Later, while being interviewed by FBI agents, he revealed the name of his US courier, Harry Gold. US agents arrested Gold, who in turn implicated David Greenglass, an American nuclear scientist who had also worked for the Manhattan Project. After Greenglass was similarly booked, he implicated Julius and Ethel Rosenberg, a couple related to him by marriage. Controversially, the Rosenbergs would be found guilty of espionage and executed. Meanwhile Fuchs was sentenced to fourteen years and served only nine. When released, he would emigrate to East Germany and live there comfortably for many decades.

Years later, Wheeler would encounter Fuchs at an international physics conference. Before walking up to him, Wheeler picked up a coffee cup and a notebook, clutching them tightly so that he would not have to shake Fuchs's hand. For mild-mannered Wheeler, refusing to shake hands was a mark of strong disapproval. He chatted politely with Fuchs for a few minutes and moved on.

CALL OF DUTY

Around the time that Truman announced the first Soviet nuclear bomb, nicknamed "Joe 1" in honor of Joseph Stalin, Wheeler had just begun a sabbatical in Europe supported by a Guggenheim Fellowship. He planned to stay in Paris and occasionally take the train up to

Copenhagen to consult with Bohr. Wheeler decided to spend the year in France partly so that his children could learn the language. It also gave him time to work on problems independently without having to participate explicitly in Bohr's group. As a hobby, he took art lessons twice a week, an interest he shared with Feyman.

By late autumn, Wheeler found himself frequently interrupted by phone calls from Teller and Harry Smyth, a Princeton colleague, urging him to get involved in new efforts at Los Alamos to develop a superbomb. With the Soviets able to build and launch fission bombs similar to those dropped on Hiroshima and Nagasaki, the United States needed far more powerful weapons to maintain nuclear superiority. Otherwise, Joseph Stalin might take advantage of American weakness to tighten his grip on Eastern Europe and expand his power to other parts of the world.

Wheeler had an extraordinarily difficult choice to make. He had his family to consider. He had a commitment to the Guggenheim Foundation that would be hard to break. There was also the matter of a graduate student, John Toll, who had taken the opportunity to reside in Paris for the academic year and collaborate with him on the problem of collisions between photons. Finally, he didn't want to disappoint Bohr.

On the other hand, he had vowed never to leave weapons development solely in the hands of government officials. Scientists had a duty to present the government with the most accurate information about capabilities. With such knowledge, tyrants such as Stalin and Adolf Hitler might be thwarted before they had the chance to cut short the lives of millions. Therefore, after making new arrangements with the Guggenheim Foundation, Bohr, Toll, and others, he decided to head back to Los Alamos. His family remained for several months in Paris, to help the kids hone their French, before joining him.

GO WEST, YOUNG PHYSICISTS

For one year, starting in the spring of 1950, Wheeler worked at the Los Alamos laboratory, deeply involved in the secret project to build a hydrogen bomb. He was assisted by Toll, who came over from Paris, and another Princeton graduate student, Kenneth W. "Ken"

Ford. Teller and Polish émigré physicist Stanislaw "Stan" Ulam, a close friend of John von Neumann, headed the efforts. They considered various schemes, finally settling on a two-stage design that became known as either the "Ulam-Teller" or "Teller-Ulam" idea, depending on who was describing it. (Teller had a tendency to minimize Ulam's contributions.) They announced the powerful blueprint for an H-bomb in a top-secret internal report, issued on March 9, 1951, titled "On Heterocatalytic Detonation I. Hydrodynamic Lenses and Radiation Mirrors." To this day, much of the report remains officially classified. Yet, as Ford relates, most of it is now common knowledge.

By that time Wheeler was already itching to leave Los Alamos and return to Princeton. Much of his impetus was family pressure. After the delights of Paris, life in an isolated military complex was a letdown. The house in the historic enclave known as Bathtub Row was fabulous; Oppy had resided there during the war. Yet Janette missed her friends, didn't like the schools, and urged John to find a way back east.

Another issue was that, aside from Teller, Ulam, and a few other well-known scientists, virtually no academic researchers could be persuaded to come out to Los Alamos for the H-bomb project. While the Manhattan Project drew from an enormous reservoir of talent, the pool seemed to have dried up. Perhaps it was a geographic issue, he speculated.

Wheeler devised an excellent scheme to continue his military research and at the same time return to New Jersey. Why not set up a kind of satellite campus of Los Alamos in the Princeton region? He could sell it to funding agencies as an "idea factory" for thermonuclear weapons. He had a hunch that such a center, located in a far more populated and accessible part of the country, would attract much more talent. At the very least, he had a number of bright students who were eager to make a contribution: Toll, Ford, and possibly others. The center would be dubbed "Project Matterhorn."

Around the same time, Feynman was experiencing his own restlessness. He was sick and tired of the frigid Ithaca winters, which seemed to last more than half the year. The omnipresent snow and ice did little to help him out of the blues. The clear Western vistas of his Los Alamos days beckoned.

Consequently, when the California Institute of Technology (Caltech) offered him an associate professorship, he jumped at the chance. Sure, he would miss Bethe and Robert Wilson, and they would miss him, but he couldn't see himself scraping ice off his car winter after winter. Southern California would be warm and sunny year-round.

Bethe decided that if anyone could fill Feynman's shoes, it would be Dyson, who would replace Feynman at Cornell starting in the fall of that year. He would stay there for a few years, before accepting a permanent position at the IAS (where, as of this writing, he now enjoys an active emeritus status).

In his zest for travel, Feynman had also accepted another offer, a ten-month visiting position in Rio de Janeiro, Brazil. Luckily, Caltech had agreed to count that time abroad as a sabbatical leave. He had visited Brazil briefly to deliver some talks and couldn't wait to experience its vibrant culture for a longer stretch.

Much to his surprise, after arriving at Caltech but before heading to Brazil, Feynman received a letter from Wheeler asking if he'd like to participate in Project Matterhorn. Wheeler continued to admire Feynman's brilliance and relished the idea of having him back at Princeton, at least for brief periods. Moreover, such a pivotal contributor to the Manhattan Project would be a tremendous asset for H-bomb development.

Wheeler began his letter to Feynman, dated March 29, 1951, with speculation about the possibility of war breaking out between the Soviet Union and the United States: "I know you plan to spend next year in Brazil. I hope world conditions will permit. They may not. My personal rough guess is at least 40 percent chance of war by September, and you undoubtedly have your own probability estimate. You may be doing some thinking about what you will do if the emergency becomes acute. Will you consider the possibility of getting in behind a full scale program of thermonuclear work at Princeton through at least to September 1952?" Disinclined to resume any military work but eager not to insult Wheeler, Feynman reiterated that he fully intended to spend his sabbatical in Brazil. Although he recognized that World War III, if it broke out, would likely disrupt his travel plans, he didn't want to change them otherwise.

As it turned out, Feynman would indeed go to Brazil; revel in its diverse, passionate culture; learn much about its educational system; and become an aficionado of Samba and other types of Brazilian music. It was an adventure he wouldn't have wanted to miss.

In April 1951, Project Matterhorn was approved and designated a site just off Princeton campus in what was called the Forrestal Building. Wheeler and his family happily moved back to their house on Battle Road. Toll and Ford joined Wheeler in the project, which would begin later that year and continue throughout 1952.

On November 1, 1952, the United States conducted the first test of a hydrogen bomb. The "Mike device," as it was called, was based on Teller and Ulam's design. A Pacific atoll called Eniwetok was chosen for the test. Mike was assembled and detonated on an island, Elugelab, which was decimated when the device exploded. The H-bomb proved more than eight hundred times more powerful than the fission bomb dropped on Hiroshima, representing a significant leap in American military might. The advantage would prove short-lived, as the Soviets would test their own H-bomb only nine months later.

Wheeler witnessed the Mike test aboard the SS *Curtis*, a ship stationed about thirty-five miles away. Even through the government-issued dark glasses he was required to wear, the brightness of the burst and the burgeoning cloud of dust it produced were unforgettable. He quickly tried to send an encoded message to Toll, Ford, and others on the Matterhorn team, letting them know of the success, but it was too cryptic for them to figure out. Fortunately, they soon learned indirectly via Teller and were elated that the bomb worked.

DERAILED

Only a few days after the Mike test, Dwight Eisenhower was elected president. It was a time of great paranoia in the US government, due to fears of Soviet spying. The Fuchs incident had cut deep and shaken up the establishment. Many military officials accused nuclear scientists of being too lax. Meanwhile, many nuclear scientists felt that the military was developing horrific new weapons too cavalierly. With finger pointing on both sides, the trust between the scientific community and the military, critical to the success of the Manhattan Project, had largely evaporated.

Many prominent scientists, such as Einstein and Bohr, had joined groups calling for international control of nuclear weapons and eventual disarmament. Their positions rattled the hawks in the US Congress and, once Eisenhower took office, in his administration. Those pressing for further bomb development suspected that Soviets were operating behind the scenes, trying to foster the peace movement to ensure their own hegemony. Opportunistic politicians, such as Senator Joseph McCarthy, promoted the idea that Communist sympathizers and former party members permeated society and occupied prominent government positions. Infamously, McCarthy began to conduct hearings investigating alleged cases of Communist infiltration. In April 1954, Oppenheimer, who had opposed developing the H-bomb, would have his security clearance revoked because of purported connections to the Communist Party. Teller testified against him. It was a dark day for physics.

Thankfully, Wheeler never faced the brutal grilling and public disgrace inflicted on Oppenheimer. Yet a security breach on a train landed him in hot water with Eisenhower. While he never lost his clearance, Wheeler did receive a presidential reprimand.

The incident happened in January 1953, as Wheeler was traveling on a late-night train from Trenton, New Jersey, down to Washington, DC. For reading material, he had brought along part of a secret document, called the Walker Report, that considered what Fuchs might have learned in Los Alamos about Teller's early ideas for a hydrogen bomb and whether the Soviets would have found such information useful in bomb building. The content included specific data about hydrogen bomb design. Wheeler was not supposed to carry such a confidential report onboard a public rail carriage, but he did so anyway to make the best use of his time.

After the train pulled into Union Station in the very early morning, it waited there for a while so that passengers could catch some sleep before departing. Wheeler woke up and realized that the secret pages were missing. After he looked frantically and couldn't find them, he contacted government agents. They impounded the carriage and searched every inch but could not find the classified material. To this day, no one knows what happened to it.

Briefed about the incident soon after he took office, Eisenhower was very upset. Smyth, who attended an Oval Office meeting with the president that focused on the breach, informed Wheeler afterward about his personal reprimand. Fortunately, by then, Project Matterhorn's major H-bomb work had already wrapped up. Wheeler's only remaining task was to complete a detailed report of what he and his team had accomplished. He had no plans to pursue further military research anyway. Rather, he had redirected his efforts back to teaching Princeton students and engaging in fundamental research. Therefore, the reprimand ultimately had little effect on his career.

WE GUESSED WRONG

Around the time of Project Matterhorn, Feynman's experience with quantum electrodynamics had made him more and more dubious of the absorber theory he had developed with Wheeler. He simply could not see how it explained known experimental results about electrons, positrons, and other particles, such as the Lamb shift. He had begun to wonder if Wheeler was experiencing the same doubts. "I wanted to know your opinion of our old theory of action at a distance," Feynman wrote to Wheeler in May 1951. "It was based on [the assumption that] electrons act only on other electrons. . . . [Yet] there is the Lamb shift in hydrogen which is supposedly due to the self-action of the electron." Indeed, by that time Feynman's contributions to quantum electrodynamics had essentially pushed the prior concept aside, like a daughter acquiring and modernizing her mother's firm. The parent theory had served its purpose well, but it had become obsolete and was due for retirement.

Much later, in his acceptance speech for the 1965 Nobel Prize in Physics, Feynman would describe how his infatuation with the absorber theory, though it proved wrong, had helped motivate his more consequential work. It led to sum over histories, which in turn motivated his diagrammatic approach to quantum electrodynamics. As Feynman would relate, "The idea [of direct interactions between electrons] seemed so obvious to me and so elegant that I fell deeply in love with it. And, like falling in love with a woman, it is only possible if you do not know much about her, so you cannot see her faults. The faults will become apparent later, but after the love is strong enough to hold you to her."

Wheeler didn't actively renounce his old ideas. Rather, he just moved on to another wild theory. Instead of viewing particles, especially electrons, as fundamental, he switched to pure geometry and energy fields. As did mathematician William Clifford in the nineteenth century and Einstein in his later years, Wheeler began to envision a universe woven entirely in the loom of geometric relationships. No sooner had he climbed down from Matterhorn than he found himself swimming in a foamy sea of cosmic speculation.

CHAPTER SIX

LIFE AS AN AMOEBA IN THE FOAMY SEA OF POSSIBILITIES

> One can imagine an intelligent amoeba with a good memory. As time progresses the amoeba is constantly splitting, each time the resulting amoebas having the same memories as the parent. . . . It would be difficult indeed to convince such an amoeba of the true situation, short of confronting him with his "other selves." The same is true if one accepts the hypothesis of the universal wave function.
>
> —Hugh Everett III, draft of PhD dissertation, Princeton University

To be normal or unconventional, that is the question. Is it better to be regular, predictable, and straightforward or bizarre, haphazard, and mercurial?

The Eisenhower era was known for conformity. Sprawling suburbs offered row after row of cookie-cutter houses. Trying to keep up with their neighbors, people flocked to department stores and grabbed the latest-model television sets. Yet what did they watch? Outrageous comedians such as Milton Berle dressing up in crazy outfits and Sid Caesar and Imogene Coca acting as goofily as possible in the silliest situations. To be popular, a show had to have at least a hint of the subversive.

Richard Feynman and John Wheeler were both aware of the contrast between the "crazy" and more conventional aspects of their lives. In Feynman's case, such nonconformity expressed itself mainly in his personal style. He never wanted to be the run-of-the-mill professor, attired in tweed, immersed in the fine details of academic committees, and making small talk over wine and cheese. That simply wasn't him. Bongos, bars, and wacky adventures were more his speed. He'd rather have tons of crazy stories up his sleeve, impressing wide-eyed listeners, than act like a traditional academic. Whenever possible, he also wanted his work to entertain him—to involve challenging calculations that presented like puzzles. Arline would have wanted him to continue to have fun.

Yet he also understood keenly the perks of settling down. He hoped someday to enjoy a real family life, like the one he had grown up with. He had the highest respect for solid individuals like Wheeler, who placed others first and had wonderful relationships with his wife and kids, as far as Feynman could tell.

Wheeler certainly appreciated the benefits of a quiet life. He had tired of bouncing from place to place like a pinball and hoped to enjoy a calmer existence in idyllic Princeton now that the H-bomb project was over. For his personal life, that would be just perfect.

His academic aspirations were a different story. He had achieved great success in describing nuclear processes and scattering. He could well have continued to publish in those fields. Doing so would have been safe; it would have been easy. However, his intellect rebelled against such predictability. His mind was drawn to the very extremes of the physical world. He loved cosmic oddities and what they said about nature.

Both men felt that life's circus was incomplete without a freak show. Freaky experiences turned Feynman on, as long as they didn't impinge upon his career as a physicist. Pounding on exotic Brazilian or African drums well into the night, picking up hitchhikers on deserted highways, visiting offbeat places, seducing pretty girls on his travels with practiced pickup lines—each lent spice to an otherwise conventional existence.

Freaky ideas intrigued Wheeler, as long as they didn't defy the fundamental principles of physics. He called his approach "radical conservative-ism." Extremes of any theory offered the best test of its validity, he felt. He loved letting his mind wander to the very limits of time, space, and perception. The tiniest things, the most basic components of the universe, the most powerful forces—all enlivened his Cirque du Soleil of thought.

Justifying his bold expeditions into the unknown, Wheeler once said, "We live on an island surrounded by a sea of ignorance. As our island of knowledge grows, so does the shore of our ignorance."

The two theorists, of course, wouldn't spend time on just any hypothesis; it had to be grounded in fact. Pseudoscience and the occult were a no go for both of them. Feynman, in a much later talk, would dub these "cargo cult science," his shorthand for a kind of primitive, untestable belief. Science could be weird, but it must be definable, observable, and reproducible. Wheeler wholeheartedly agreed.

THE WARP AND WOOF OF HISTORY

The classical and the quantum illustrate the stark contrast between predictability and weirdness. Classical mechanics maps out the lives of particles so precisely that they might as well be grey-suited businessmen taking the same route to the office each day. Quantum mechanics, on the other hand, is capricious, steering particles one way or another like a finicky bouncer guarding the entrance to a club with arbitrary standards. In public presentations, Feynman would emphasize how unintuitive quantum physics really was. For example, in a lecture series titled *The Character of Physical Law*, Feynman said, "I think I can safely say that nobody understands quantum mechanics."

Yet, through the idea of sum over histories, Feynman wonderfully harmonized those extremes. It brilliantly showed how, among all the outlandish possibilities quantum physics might conjure, the least action principle mandated that the staid classical path is the likeliest. Veer into the subatomic realm, however, and more and more quantum weirdness enters the picture, including corrections represented by various types of Feynman diagrams.

Believing that Feynman's melding made much more sense than either classical or quantum mechanics on its own, Wheeler wanted to spread the word. He knew that many great minds were already working on the implications of quantum electrodynamics. Therefore, he tacitly decided to steer clear of that subject and search for ways to apply the sum-over-histories concept to other unexplored frontiers, such as gravitation.

Wheeler's neighbor across the street was a weaver. The weaver's craft—the "warp and woof" actions of the loom—fascinated him. As his daughter Alison recalled, he became intrigued by the idea of stitching together reality from the threads of different possibilities. He began to consider the idea that sum over histories applied to the universe itself. "My father spoke about the warp and woof of history," Alison related. "You have the streams and the cross-currents coming to enrich the history."

To weave together such a vision, Wheeler would need to learn to use two different looms in conjunction. The first was Feynman's loom, sum over histories, which he knew well already. It was sparkling new and primed for use. The second was Albert Einstein's loom, the general theory of relativity, which guides how spacetime's fabric is stitched together. As Einstein had constructed the theory in 1915, many physicists considered it obsolete. Seeing its potential, Wheeler would dust it off, oil its gears, and get it working even better than new.

A PAUCITY OF POSSIBILITIES

By the late 1940s and early 1950s, virtually no young researchers in the United States were interested in general relativity. There was Peter Bergmann, who had worked with Einstein and written one of the few textbooks on the subject. There was Bryce DeWitt, who had read Bergmann's textbook and wanted to bring the theory into line with Julian Schwinger's methods for quantum electrodynamics. Aside from those scattered exceptions, most graduate students and young professors wouldn't touch the subject with a ten-parsec pole (a bit of an exaggeration, but you get the idea). Many major journals, such as *Physical Review*, similarly eschewed papers on the topic.

One might speculate about the reasons for such avoidance. Einstein himself was elderly, a reminder of how long it had been since his 1915 theory had made headlines. That had happened in 1919, with solar eclipse measurements confirming one of its key predictions.

Aside from data taken during other solar eclipses, little in the way of new experimental results inspired continued interest in the subject. Einstein didn't even pursue it, focused as he was on unified field theories that modified general relativity to bring electromagnetism into the picture.

The major application of general relativity had been in cosmology, the study of the universe. That subject encompassed a few well-known solutions of Einstein's equations, found by making simplifying assumptions, such as the uniformity of space. The key breakthrough was Edwin Hubble's 1929 discovery that all the galaxies in the cosmos (save our nearest neighbors, which are gravitationally bound to us more tightly) were moving away from each other, generally interpreted as proof of the expansion of space.

Two schools of thought debated what spatial expansion meant. The first, headed by George Gamow and based on the ideas of Alexander Friedmann, George Lemaitre, and others, advocated that the universe must have had a fiery beginning, when all its matter and energy were concentrated in a tiny region. Gamow pointed to two pieces of evidence for this. The first was the Hubble expansion. The second was the buildup of higher elements, such as helium, from simple hydrogen. The bulk of helium observed today could have been created only during the fiery conditions of a compact early universe. Stars couldn't have generated enough helium to produce the current amount in the cosmos. Working with Ralph Alpher, Gamow published that hypothesis in 1948. Playing on the first three letters of the Greek alphabet—alpha, beta, and gamma—Gamow listed Bethe's name between Alpher's and his own, even though Bethe hadn't directly played a part in the research. Hence, with Alpher, Bethe, and Gamow as the listed authors, it became known as the "alpha-beta-gamma" theory.

Three astronomers, Fred Hoyle, Hermann Bondi, and Thomas "Tommy" Gold, led the opposing camp. Dubbing the model of Gamow and Alpher the "Big Bang," Hoyle ridiculed the concept that all the matter and energy in the universe emerged instantly out of pure nothingness. Rather, his "Steady State" cosmology held that new material must pepper the cosmos gradually, in minuscule amounts, over the eons. As the universe expands, the virgin matter slowly coalesces into fledgling galaxies that eventually repopulate the gaps between the old. Hence the cosmos maintains approximately the same appearance over time.

Both the Big Bang and Steady State models obey the cosmological principle, which states that space and its material are approximately uniform everywhere in all directions. In other words, no part of the universe looks completely different from the rest. As a result, they are modeled in general relativity by simple solutions of Einstein's equations that are both isotropic (identical in all directions) and homogeneous (similar everywhere). Only a handful of solutions meet those requirements.

As Bondi and Gold emphasized, the Steady State concept also conforms to what they called the "perfect cosmological principle." An even stricter requirement, it mandates that the universe, as a whole, looks approximately the same throughout all time as well as all space. Because new galaxies arise whenever there are gaps, the universe is continually injected with a kind of cosmological collagen and ages as little as Dorian Gray. Like the cosmological principle, the perfect cosmological principle limits the range of possible solutions to Einstein's equations.

With only a bit of general relativity needed to understand known cosmological solutions and few applications otherwise, no wonder few physicists were flocking to general relativity in the early 1950s. Many viewed it simply as a "playground for mathematicians." Nevertheless, a fresh look at a 1939 paper by Robert Oppenheimer and thoughts about the fundamental ingredients of nature inspired Wheeler to buck the trend and help revitalize the field.

THE FRINGES OF GENERAL RELATIVITY

In early 1952, with Project Matterhorn in full throttle, Wheeler's thoughts veered momentarily to fundamental questions about the universe. In revisiting his nuclear work with Niels Bohr, he likely took another look at the September 1, 1939, issue of *Physical Review,* where their key article was published. In the same issue was another article, written by Oppenheimer and his student Hartland Snyder, titled "Continued Gravitational Contraction," about a possible scenario for the final states of massive stars.

Strangely enough, Oppenheimer and Snyder had predicted that under certain circumstances, heavy stars that had exhausted their nuclear fuel and become unable to expel most of their material would catastrophically collapse in an unstoppable way. Within a finite period, these stellar giants would implode into infinitely dense states, known as "singularities." Their gravitational oomph would be such that nothing, not even light, could escape from small spherical regions around their centers. Therefore, no one on the outside could peer in and record what was happening inside. Wheeler would later refer to such objects as "black holes."

At that point, however, he didn't quite believe the Oppenheimer-Snyder result. He thought that some type of mechanism—perhaps a quantum process or other type of smoothing out—would halt the collapse before it resulted in a singularity. Singularities seemed the fly in the ointment for other physical theories too, such as the self-energy of the classical electron. Perhaps resolving the gravitational question would reveal something fundamental about how nature avoids singularities. To find a reasonable alternative, he'd need to study general relativity thoroughly until he had mastered the subject.

The best way to learn a field was to teach it, Wheeler had found. He had acquired the habit of assembling meticulous lecture notes for each course, which could double as an excellent resource whenever he continued to research a subject. Often in his notebooks, he scattered speculations among his course notes. He might ask those questions of his students, consider them himself, or both. Learning begets teaching, which begets more learning, in a marvelous spiral of rising knowledge.

Wheeler asked his department to allow him to design and teach Princeton's first full-year course on both special and general relativity. On May 6, he received permission to offer it the following academic year. He celebrated the news by writing "Relativity I" on the cover of a fresh notebook, noting the date and time, and starting to plan the course material. He also noted that he hoped to develop a book on the subject some day—a promise to himself that he'd fulfill splendidly, years later, with the definitive text *Gravitation* (coauthored with Charles Misner and Kip Thorne).

SUPERFLUIDS AND A NOT-SO-SUPER MARRIAGE

Meanwhile Feynman was preparing for a different kind of milestone: his second wedding. Shortly before he had left for his sabbatical in Brazil, he had met a young woman at Cornell named Mary Louise "Mary Lou" Bell. She hailed from the rural Midwest and had an encyclopedic knowledge of art history. On the face of it, aside from an interest in art (which he'd developed when he started dating Arline), they didn't have much in common. However, while in Brazil and pining for a real relationship—rather than flings with flight attendants—he fondly remembered some of their more interesting conversations about ancient relics and such. Unconventionally and impulsively, he proposed to her by mail. She agreed to marry him and live together in Southern California. They wed in June 1952 and took up residence in Altadena, not too far from Pasadena, where Caltech is located. Honeymooning in Mexico, Feynman learned much about Mayan art and culture, including its "calendar round" concept of cyclical time.

Other changes were afoot. In separating himself from Hans Bethe and the Cornell group, Feynman gained the opportunity to reassess his research direction. While he derived an enormous sense of accomplishment from his quantum electrodynamics work, he harbored major doubts about the methods used in renormalization. At some point, he felt, that process needed a fundamental overhaul to nail down some of its arbitrary aspects. In the meanwhile, perhaps it was time to do something completely different.

In his book *QED*, published decades later, Feynman would write, "The shell game that we play . . . is technically called 'renormalization.' But no matter how clever the word, it is what I call a dippy process. Having to resort to such hocus-pocus has prevented us from proving that quantum electrodynamics is mathematically self-consistent. . . . I suspect that renormalization is not mathematically legitimate."

During his early years at Caltech, Feynman investigated superfluids, substances such as liquid helium that never solidify, no matter how low the temperature drops. Basing his research on the work of Soviet physicist Lev Landau, he considered why quantum mechanics might prevent such systems from lowering their energies and becoming solid. His superfluidity work proved critical to the emerging field.

Feynman also investigated superconductivity, another low-temperature phenomenon, in which materials lose their electrical resistance below a certain critical temperature. Currents, once set in motion, just keep flowing. Similarly, Feynman believed, a quantum mechanism must exist to maintain currents in such states of constant flow. John Bardeen, Leon Cooper, and J. Robert Schrieffer would discover the specific mechanism soon thereafter in research for which they won the Nobel Prize. It frustrated Feynman that he didn't arrive at the answer himself, despite his important contributions to superfluidity.

Invited by Wheeler, Feynman presented a talk on the superfluidity of liquid helium at a conference in Japan in September 1953. He enjoyed learning and practicing some Japanese phrases. Intrigued by Japanese rituals, he stayed at a traditional hotel and decided to partake of a customary bath. Embarrassingly, he walked into the wrong room and accidentally barged in on the renowned physicist Hideki Yukawa soaking in a tub.

Feynman's marriage was cooling off even faster than the frigid theories he was exploring. He and Mary Lou were a complete mismatch. His trademark casual style was a no go. She asked him to dress up for work in a jacket and tie instead of his usual shirtsleeves. His colleagues found her unpleasant. She wanted his undivided attention, even at times when he preferred to retreat into thought. Soon he was tuning her out altogether. His drumming drove her crazy, and his calculations baffled her to no end. Within a few years, she would be pressing him for divorce.

At their divorce trial, in July 1956, Mary Lou would unleash her built-up frustration. "He begins working calculus problems in his head as soon as he awakens," she testified. "I couldn't talk to him because he would say I was interrupting his work." Moreover, his drums made a "terrific noise."

The court ordered Feynman to pay her a cash sum, along with regular alimony payments. While forced to give her some of his property, newspaper accounts of the trial reported with amusement, he got to keep his drums.

ON EINSTEIN'S TURF

Wheeler taught his first yearlong course in relativity in fall 1952 and spring 1953, with the latter semester focusing primarily on general relativity. Among the highlights of the spring session was a class trip on May 16 to Einstein's house on Mercer Street for tea and discussion. There, the eight graduate students in attendance had the unprecedented chance to ask the theory's designer any questions on their minds. The universe was the limit. Einstein captivated the students with his thoughts on the expansion of space, his critique of quantum mechanics, and other matters. He even answered a somewhat morbid question about what would happen to his house after his death. "This house will never become a place of pilgrimage where people come to see the bones of the saint," Einstein responded.

When they were getting ready to leave, Wheeler asked Einstein if he had any advice for budding young physicists. "Who am I to say?" he replied.

Albert Einstein, Hideki Yukawa, and John Wheeler strolling together through Marquand Park, Princeton, in 1954 (*Source:* Photograph by Wallace Litwin and Josef Kringold, courtesy AIP Emilio Segrè Visual Archives, Wheeler Collection).

In preparing his notes for the course, Wheeler got to know general relativity inside and out. He found a succinct way to describe it, which he would later use in his classic textbook: "Space acts on matter, telling it how to move. In turn, matter reacts back on space, telling it how to curve."

In general relativity, the Earth moves in an elliptical path around the sun not because of forces acting over a distance but because of the local warping of space due to the sun's mass. If the sun suddenly disappeared, the space around it would soon flatten out due to gravitational ripples traveling at the speed of light. Once the Earth's spatial region became pancake flat, it would start traveling in a straight line rather than a curve.

You don't need mass to create twists or dents in space. Energy fields do just fine because in Einstein's theories energy and mass are perfectly equivalent. Place a glob of energy in any region, and it starts to curve. Even gravitational energy, due to spatial ripples, might create its own warping. Therefore, one might imagine a feedback loop in which ripples generate their own ripples. In other words, geometry engenders itself, without the need for matter.

General relativity soon became Wheeler's canvas. He aspired to craft anything in physics out of warped spaces and energetic fields. He dropped the concept of "everything is particles" and adopted "everything is fields." It was a complete turnabout. Once he thought fields were an illusion; now he started to feel the same about material things. Once he believed in action at a distance; now all happened locally. Seeing how the brave new world of geometry and fields panned out would be an exciting adventure.

A DIET OF WORMS

A 1935 paper by Einstein and his assistant at the Institute for Advanced Study (IAS), Nathan Rosen, explored an intriguing kind of general relativistic geometry in which two spacetime sheets were connected by a narrow passage, which became known as the "Einstein-Rosen bridge." The figure they constructed looked something like an hourglass, with the upper half being one region, the lower, another, and the throat acting as the connecting bridge. Einstein and

Rosen tried to use such geometric constructions as stand-ins for elementary particles such as electrons.

Wheeler took the same kind of structure and rechristened it a "wormhole." He imagined it as the surface of an apple, representing conventional space, with a worm (energy causing extreme warping) gobbling out spatial shortcuts. The resulting wormholes changed the topology (mathematics of connections) of space. Once such wormholes appeared, other energy fields could thread through them, allowing them to skip from one point to another as if by magic.

Once Wheeler started contemplating all the things wormholes could do, he had a stunning insight. Suppose a cluster of electromagnetic field lines fell into the mouth of a wormhole. They would appear to converge at a point, mimicking what a negative charge, such as an electron, would do. Those gobbled lines, after traveling through the wormhole's throat, would emerge from a second mouth. They would appear to emanate from another point, as if emitted by a positive point charge, such as a positron. Hence, the wormhole seemed to create a pair of opposite charges. Charge appeared to arise from sheer nothingness, an effect he dubbed "charge without charge."

Wheeler investigated another structure: a self-contained clump of energy he called a "geon." He wondered what would happen if an electromagnetic field was shaped in such a way—as a sphere, donut, or other configuration—that its own gravity kept it stuck together indefinitely. Such gravitational putty would act, for all practical purposes, like a particle. Following Einstein's dictum that energy and mass are freely convertible, it would even have mass. Wheeler called this notion "mass without mass."

Wheeler began to dream of a new kind of physics built up from wormholes, geons, and other geometric constructs, along with energetic fields. Matter would simply be an illusion in the emerging science of "geometrodynamics." He would need to do a lot of calculations, he realized, to see if his radical new model worked.

General relativity is notoriously difficult to solve exactly, except in simple cases. Yet in Project Matterhorn he learned that computational techniques could go a long way. He just needed clever students—like John Toll and Kenneth Ford in his H-bomb work—to master the techniques and get the geometrodynamics program off the ground. Spectacular new vistas awaited, and he was excited.

CHARLIE AND THE GEOMETRY FACTORY

Brilliant theoreticians come along once in a blue moon. When Wheeler met Charles "Charlie" Misner, he was absolutely delighted to find a young, mathematically savvy graduate student as passionate about general relativity as he was. Misner had mastered the field in his undergraduate work at Notre Dame, along with acquiring an incredible arsenal of powerful mathematical techniques, including topology. He entered Princeton's graduate program in the fall semester of 1952. Like Feynman, he resided at first in the Graduate College and grew familiar with the walk over to the Palmer Lab complex. In his first year, while taking courses and working on a radioactive decay project with physicist Arthur Wightman, he encountered Wheeler several times at Palmer, taking the opportunity to chat with him about general relativity. Consequently, as Wheeler began to conjure up wormholes, geons, and other oddities of geometrodynamics, Misner stepped into the project on the ground floor. He had the theoretical background to understand it in a profound way that highly impressed Wheeler. Soon they contemplated climbing its lofty heights together, with Wheeler serving as Misner's PhD supervisor.

In his discussions with Misner, Wheeler developed a conceptual model of quantum gravity that he called "quantum foam," also known as "spacetime foam." He imagined a kind of ultrapowerful microscope magnifying nature's fabric and revealing what happened at its tiniest scale, called the "Planck length," about 6×10^{-34} inches. A trillion trillion objects of such size, placed end to end, still wouldn't span an atom. Clearly, for such minute distances, quantum rules would need to come into play. Spacetime would be "frothy," writhing to the pulse of random quantum fluctuations. Wormholes and other connected structures would spontaneously arise and just as quickly dissipate, forming and popping like fleeting bubbles in the foam.

Understanding how classical reality emerged from such chaotic froth would require a sort of optimization process that singled out the ordered cosmos that we see today. Wheeler wondered if a kind of sum over histories applied to spacetime itself would do the trick. Among all the possible evolutions of the universe, perhaps Feynman's methods could identify the one taking the optimum path through an abstract space representing all geometric possibilities. The result

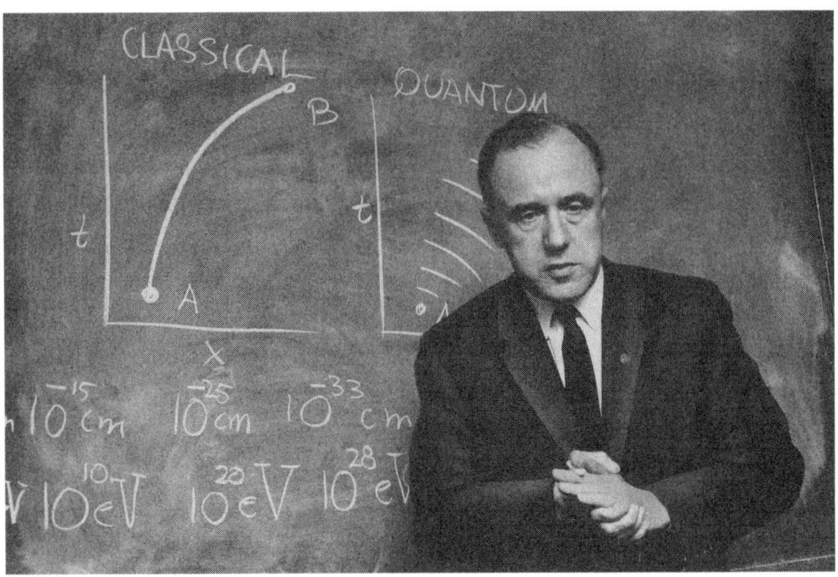

John Wheeler at Princeton in 1967, explaining the difference between classical and quantum processes using Feynman's sum-over-histories method (*Source: New York Times*/Redux).

would be a beautiful link between classical and quantum gravitation that explained how order arose from random quantum fluctuations.

Wheeler's students often teased him for being overambitious. He freely admitted that procrastination sometimes delayed him from fulfilling his visions. His dreams, moreover, constantly took him to new places. The typical Wheeler move that invited such teasing would be to come up with a ridiculously short time line for a far-reaching, multipart project and to only get around to part one before switching to something else. We've seen how that happened with the first Wheeler-Feynman absorber paper.

With Misner, Wheeler was similarly ludicrously optimistic. He assigned Misner the task of applying Feynman's methods to geometrodynamics to produce a quantum theory of gravity. It wouldn't take long, Wheeler emphasized, without really thinking about the challenges. "This sum over histories idea for doing quantum gravity, you can probably turn that out in six months," he assured Misner.

Misner found the sum-over-histories concept fascinating. He saw as its essential lesson that "reality proceeds by an awareness of all the possibilities before you arrive at actuality." Nevertheless, it seemed that a full classical (nonquantum) description of connections between electromagnetism and gravitation was required before the quantum version could be pursued.

Misner got to work and started developing mathematical theorems related to general relativity. At the time, Wheeler didn't know the pertinent mathematical literature very well and could not provide much guidance in that area. However, Peter Bergmann, who had a look at the developing project, pointed out what had been done before. Some of Misner's initial work, Bergmann noted, was similar to results published in 1925 by mathematical physicist George Rainich. To avoid duplicating others' efforts, Misner decided to focus his thesis on applying sum over histories to general relativity, as ambitious as that sounded. During his time at Princeton, he would make great progress in formulating the problem and outlining further steps toward developing a quantum theory of gravity using Feynman's methods. He would come to realize, however, that completing such a theory would not be so easy. His PhD thesis, "Outline of Feynman Quantization of General Relativity," would thereby offer a solid starting point rather than a culmination.

Thanks to exceptional students such as Feynman and Misner, Wheeler began to see himself as launching a new school of thought, in the manner that Bohr did in Copenhagen. He certainly had enough projects related to general relativity and the quantum to go around. He hoped that the best of his students would remain long-term collaborators, friends, and confidantes. Indeed, many of his former students—including Ford, Toll, and others—would stay in touch as they rose in their careers and collaborate with him on a variety of academic endeavors.

Wheeler loved working with young thinkers so much that he opened the gates to many whom other professors might consider unconventional. A prime example was Peter Putnam, a student interested in the intersection of philosophy, psychology, and physics. An extremely thoughtful but shy and lonely young man, he attended Princeton as an undergraduate shortly after his brother died in combat during World War II. Wheeler felt very close to Putnam because of their mutual philosophical interests and would later take him on as a graduate student. They had many deep conversations about the role of subjective experiences in shaping reality. No doubt, part of their camaraderie stemmed from the loss of a brother killed in the war. Although Putnam would ultimately leave physics and languish in poverty, their intense discussions shaped Wheeler's later views on connections between conscious awareness and the quantum world.

THE FOUR ACES

At the Graduate College, Misner had acquired a close group of friends who shared a love of mathematics and a passion for poker and ping-pong. The four young men enjoyed many fun times together and kept in touch even after their graduate school days. Canadian-born Hale Trotter would end up remaining in Princeton as a math professor, eventually becoming department chair. Harvey Arnold was a budding young statistician. Hugh Everett III started out in the branch of mathematics called "game theory," deeply related to probability theory. He also took physics courses, including in electromagnetism and quantum mechanics. For the latter, the textbook was John von Neumann's classic work. The brainy quartet often hung out in Everett's room, pouring glasses of sherry or mixing cocktails, playing games, and holding intellectual discussions late into the evening.

Everett was a longtime aficionado of Einstein's theories. When he was twelve, he had written Einstein a fan letter, asking what holds the universe together. Einstein sent him a brief, friendly reply, gently teasing him for his dogged curiosity.

Einstein's seventy-fifth birthday took place on March 14, 1954. Shortly before the celebration, a committee headed by Oppenheimer met to decide the recipient of the Albert Einstein Award, a special prize in his honor. The winner was Feynman, who received $15,000 and a gold medal. The *New York Times* devoted an article to Feynman's achievement, noting that the award was the "highest of its kind and next only to the Nobel Prize."

On April 14, Einstein gave a special lecture at Palmer as the invited guest of Wheeler's general relativity class, which he was teaching for the second time. Those organizing the talk, including physics graduate student (and later discoverer of the color charge in the strong force) Oscar Greenberg, who introduced Einstein, kept it a closely guarded secret, fearing a mass invasion by those who just wanted to see the great man. Naturally, Misner knew about the talk and wouldn't miss it for the world. He later recalled that Everett also attended.

Einstein emphasized at the talk his belief that while quantum mechanics had been enormously successful in predicting the outcomes of experiment after experiment, it had logical gaps. He found it ridiculous, for example, that human observers and the process of measurement remained an integral part of the theory. If people are required to trigger the collapse of a wave function down to states representing particular measured values, he wondered, why couldn't a mouse do the trick? The whole process of quantum measurement required objective, mechanistic, and authoritative recasting.

Though the main thrust of his research was game theory, Everett—inspired by his first-year quantum class, his readings of von Neumann, and Einstein's talk—started thinking about the challenges of the quantum measurement problem. Coincidentally, just around this time Wheeler had started to send out feelers to see who among the graduate students might be interested in projects related to general relativity and its quantization.

In the fall of that year, Bohr spent a semester in residency at the IAS to consult with Wheeler, Oppenheimer, Eugene Wigner, and others. He brought along his young assistant, Aage Petersen. On November 16, he gave a talk at the Graduate College, which Misner, Everett, and Petersen attended. One of the main issues he addressed was quantum measurement theory.

Just as Einstein hadn't budged, since the mid-1920s, in his opposition to quantum dice rolling, Bohr hadn't strayed from his unique interpretation of quantum physics: complementarity. Bohr emphasized that quantum mechanics is a kind of black box. The answers we get out of it depend on the types of measurements we perform. If we conduct an experiment to test a system's particle properties, we get a particle-like answer. If we switch to a wave type of experiment, a wavelike answer pops out. Full knowledge of the subatomic world will never be achieved, he believed; there will always be quantum mysteries. Like devotees of Eastern mysticism, we just need to accept that some of nature's riddles have no answer. Bohr's inscrutable manner of speaking—quiet mumbling—rendered his pronouncements even more enigmatic.

Graduate students conversing with Niels Bohr on his 1954 visit to Princeton. Left to right: Charles Misner, Hale Trotter, Niels Bohr, Hugh Everett III, and David Harrison (*Source:* Photograph by Alan Richards, courtesy AIP Emilio Segrè Visual Archives).

While greatly admiring Einstein and Bohr, the quantum physics community had largely moved on to more practical interpretations of quantum measurement. Von Neumann's idea, cohesively expressed in his textbook, offered the best expression of what had come to be called the "Copenhagen interpretation." When taking quantum measurement, you single out a particular physical parameter, called an "observable." For example, if you rig up a way to pin down a particle's location, position is the observable. Before the measurement the quantum system consists of a combination of possibilities—for example, a mixture of a certain amount of one position state, a certain amount of another, and so forth—called a "superposition of states." This amalgamation evolves continuously according to the Schrödinger equation. However, the instant a position measurement is taken, everything changes. The system collapses randomly down to one of the position states, like an unstable house of cards toppling in an arbitrary direction.

Everett didn't like any of the existing quantum interpretations. Each seemed arbitrary and subjective. One evening, fortified with copious sherry, he blurted out his feelings to Petersen, who was staying at the Graduate College. Quantum physics sorely needed an objective explanation, he argued. The idea of observables, based on what is being measured, seemed absurd. Why should a human experimenter's choice affect the doings of the particle world?

Petersen felt obliged to stick up for his mentor. In his view, quantum measurement was pretty much a settled matter. While Bohr's complementarity supplied the philosophical underpinnings, more detailed interpretations such as von Neumann's showed exactly how to calculate experimental outcomes. There were so many exciting avenues of physics to explore. Why reinvent the wheel by calling into question a theory as successful as quantum mechanics?

Misner witnessed the debate with great interest. There would be more such discussions as the semester progressed. He could appreciate both sides of the argument. On the one hand, he sympathized with Everett's doubts in the face of Petersen's dogmatic defense of the mainstream view. "Hugh thought that Petersen's interpretation was intolerable," he recounted.

Misner agreed with Everett that it was odd that the Schrödinger equation worked for continuous evolution but not to explain measurement. "That seems a strange attitude for a fundamental law of physics," he recalled thinking.

On the other hand, Misner had more immediate issues to address. He was hard at work grappling with geometrodynamics, with a view to the quantum. Applying Feynman's methods to gravitation, though challenging, offered a more tangible problem than rethinking the entirety of quantum measurement theory in some quixotic quest for objectivity.

THE WAVE FUNCTION OF THE UNIVERSE

In their pursuit of quantum gravity, Misner and Wheeler soon realized that the choice of observables (basic physical quantities to be measured—the equivalent of position, momentum, energy, and so forth, in describing simple particles) wasn't obvious. General relativity, in the original form Einstein developed, placed space and time on similar footing. One might easily be transformed into the other, like ice and water. However, in any type of realistic observation we notice how something changes spatially over particular time intervals—such as the universe expanding throughout the eons. Therefore, the process of measurement naturally separates space and time, requiring that the four-dimensional block of spacetime be carved into three-dimensional slices over time. The slicing couldn't just be arbitrary; it needed to respect Einsteinian dynamics. Physicists Richard "Dick" Arnowitt and Stanley Deser, along with Misner, would find the solution to that formidable problem several years later, dubbed the "ADM formalism."

A more philosophical issue was separating the observer from the observed. If the entire universe was the system under study, how could an independent observation take place? Naturally, no one could leave the universe to look back and measure it. On the other hand, anyone inside the universe would be part of the system. Seeing how the concept of wave function collapse might work, without referring to an external observer, would be challenging. Developing a way to ascertain a quantum measurement without such a collapse would be equally daunting.

One day, Everett, who had been thinking deeply about alternatives to the Copenhagen interpretation, approached Wheeler with a revolutionary new hypothesis: suppose there was no collapse. Suppose that the wave function of each quantum system, and indeed of the universe as a whole, kept evolving smoothly according to the Schrödinger equation, the Dirac equation, and all continuous ways of describing quantum systems. If there was no collapse, there would be no need for independent observers. Therefore a wave function of the universe might be defined unequivocally.

There's only one catch, and it's a doozy. Imagine for a basic quantum system, such as an atom, that a scientist did make an observation. Such measurements happen all the time and, if conducted correctly, generally result in a single answer, not an array. In such cases, the observation, in the instant it happened, instead of initiating a collapse would cause the universe itself to bifurcate into multiple possibilities. Each branch would represent a different outcome, an alternative reality, so to speak.

Misner was intrigued by, but dubious of, Everett's hypothesis, which he had seen under development. As he recalled, "My first and general reaction was that I didn't like the conclusions but I respected Hugh's ability to argue logically. I certainly didn't like the Bohr hypothesis."

Wheeler, however, was genuinely excited. He had been pressing the graduate students for a way to define the wave function of the universe, and now Everett seemed to have one possible answer. It was, at the very least, worth a closer look, given the dearth of logically consistent alternatives. After discussions between the two of them, Everett decided to develop a PhD thesis on the topic with Wheeler as his supervisor.

THE TROUBLE WITH GEONS

Meanwhile, Wheeler continued to play with wormholes and geons, in their various forms, like an excited child with a construction set. With the latter, he had run into a serious roadblock. He wanted to make geons a key component of the particle world. Yet in his calculations he determined the minimum size of a doughnut-shaped classical geon to be around the size of the sun. It would have the mass of about a million suns, hardly qualifying it as an elementary particle. Nonetheless, he pressed on, thinking the concept far too intriguing to simply abandon. It was also conceivable that quantum corrections would help slim down its mass and size.

While on a brief trip to Europe, Wheeler had written to Einstein seeking his advice on geons. Einstein indicated that they should discuss the matter upon Wheeler's return. In October 1954 they had a friendly chat by phone. Einstein offered his gut reaction to the geon idea. While they likely constituted actual solutions to his equations of general relativity, they were probably highly unstable. Because gravitation is so weak compared to other forces, it would be hard to assemble a stable configuration of energy fields via their gravitational interaction alone. Stable astronomical bodies, such as stars and planets, are chock-full of mass, not just amorphous fields. After Wheeler performed more calculations, he determined that Einstein had been right on both counts. Wormholes, of the sort he proposed, similarly turned out to be unstable. Each would evaporate if overloaded with matter or energy, like an overinflated tire bursting on the slightest impact. Wheeler still pressed on with the idea, hoping he could somehow find stable solutions, like persistent waves in a turbulent ocean. He started writing up a long paper about geons.

During the spring semester of 1955, Wheeler was excited to teach a course on advanced quantum mechanics. As usual, he planned out the lecture notes carefully. He made sure to include ample references to Feynman's sum-over-histories method. According to his enrollment sheet, Misner and Everett both audited the course. The two friends, who had moved off campus and become roommates, were hard at work on their respective thesis projects. While Misner's research connected much more closely with the course material, Everett had the chance to learn more about Wheeler's perspective, which had been less familiar to him.

When Einstein died on April 18, Wheeler lost one of his heroes. The two had grown closer in Einstein's final years, especially when Wheeler decided to plunge into general relativity. Thanks to Wheeler and others, interest in that field had started to grow. Ironically, the decades after Einstein's death would widely be viewed as the subject's golden age.

Sometime in early fall, Wheeler sent Feynman an advanced copy of his seventeen-page treatise on geons. Apparently, he wanted Feynman to take a look and think about possible ways to add quantum corrections to the theory. With his (soon-to-be-ex) wife Mary Lou nagging him incessantly about his drumming, his calculations, and whatever else ticked her off, Wheeler's bugging him about geons must have been a pleasant diversion.

Feynman replied on October 4 with a brief, seat-of-the-pants sum-over-histories analysis of a possible first-order quantum correction to geons, akin to what he had done with the Lamb shift. He did so as a kind of fun mental exercise, unconvinced that realistic, stable geons were possible (in fact they weren't). He asked Wheeler for more proof of the concept.

Feynman's sneak peek at Wheeler's bizarre new scheme got him thinking about gravitation and why it differs radically from the other forces. Gravity is so much weaker than electromagnetism that any structures built from it must be enormous. An atom built with gravitational rather than electromagnetic glue would be astronomical in girth. Before even thinking about quantizing gravitation, he concluded, one must address the deeper question of why the force's weakness stands out like a sore thumb.

Feynman would later write in an article about quantum gravity, "There's a certain irrationality to any work in gravitation . . . shown . . . in the absurd creations of Prof. Wheeler and other such things, because the dimensions are so peculiar." An essay by Bryce DeWitt about practical uses for gravitation, awarded first prize by the Gravity Research Foundation in 1953, made a similar point. Any devices constructed exclusively from the gravitational force would need to be planetary in scale because of that interaction's relative feebleness. Assembling such astronomically large configurations would exceed our present capabilities and require a far more advanced civilization.

In 1955 wealthy industrialist Agnew H. Bahnson Jr. was so impressed by the prize-winning essay that he set up a center, the Institute of Field Physics, associated with the University of North Carolina (UNC), Chapel Hill, with DeWitt as professor and director of research. Cécile DeWitt-Morette, though an extraordinary researcher in her own right, was appointed to a nonpermanent position as visiting research professor. The ostensible purpose of the center was to investigate the possibility of antigravity technology for flight, but DeWitt broadened its mission to explore the properties of gravitation in general. With its expanded focus and ties to the university, Wheeler, Feynman, Oppenheimer, Toll, and Freeman Dyson all expressed support for the institute. Along with Princeton and Syracuse (under Bergmann), it became an important hub for the study of gravitation in both its classical version and quantum speculations.

The perplexing issue identified by Feynman, DeWitt, and others—the strength imbalance between gravitation and the other forces of nature—remains unsolved today. Given that many physicists believe that all the natural interactions were of the same strength at the time of the Big Bang, the peculiar weakness of gravity persists as one of the deepest riddles of science.

WHEN REALITY SPLITS

By the fall of 1955, Everett had thought through many aspects of his universal wave function idea, as the concept of no collapse became known, and was ready to share them with Wheeler. He handed Wheeler a few different mini-reports, including one titled "Probability in Wave Mechanics," which was rich in descriptive language and apt analogies. Based on Wheeler's comments, he combined these into a draft of his thesis.

Unlike the jolting concept of wave function collapse, Everett's scheme was as smooth as silk. A measurement would never lead to a discontinuity. Rather, the interaction between the observer and the system being measured would seamlessly result in a certain state. To take into account cases in which a quantum measurement might lead to one of several different results, Everett argued that each would be a valid end state, achieved through a branching of reality. The observer would split too, into different, nearly identical versions of himself, distinguished only by the result of the measurement each witnessed. None of the copies would know about the others, as they would live the rest of their lives in wholly different limbs of time. Everett wrote, "As soon as the observation is performed, the composite state is split into a superposition for which each element describes a different object-system state and an observer with (different) knowledge of it. Only the totality of these observer states, with their diverse knowledge, contains complete information about the original object-system state—but there is no possible communication between the observers described by these separate states."

Take, for instance, the famous conundrum described by Erwin Schrödinger in which a cat is placed in a closed box with a vial of poison, a Geiger counter, and a radioactive sample with a 50 percent chance of decaying in a given time. The system is rigged so that if the Geiger counter registers decay, the vial is smashed, the poison released, and the cat killed. On the other hand, if the Geiger counter isn't triggered, the vial remains intact, and the feline is spared. To point out the absurdity of the Copenhagen interpretation, Schrödinger argued that it implied the cat would be in a zombielike superposition of dead and alive until the box was opened and an observer measured the system.

Everett's interpretation made a completely different prediction. Once the system was set up such that the cat's fate was entangled with that of the sample, reality would bifurcate. In one branch, the sample would decay, the counter would click, the cat would be poisoned, and the observer would be in tears. In the other, the cat would live, and the observer would rejoice. The happy and sad versions of the scientist, replicated—except for the experimental outcome—at the moment of measurement, would know nothing of each other and have no possibility of comparing notes. They'd live in two slightly different alternative branches of reality—separated in an abstract realm of all possibilities. There would be no "bump" when reality split, so neither would find anything unusual. Without any collapse of the wave function, it would continue to flow as steadily as a river that divides into two streams.

While Wheeler liked the idea of a universal wave function and Everett's general approach to avoiding collapse, he was uncomfortable at that point with references to observers' experiences of reality. Consciousness was not the province of physics, he felt at the time (his views on that matter would become more flexible as he got older). As he didn't want Everett's opus to irritate the other members of the thesis defense committee, he struck any references to splitting, perception, and so forth. Wheeler also didn't want to alarm Bohr, who he hoped would appreciate the work. For Bohr to understand it and think of it as a step forward, Everett would need to tread carefully.

For example, a brilliant analogy by Everett, about amoebas dividing in two, ended up on the cutting-room floor. He imagined an intelligent one-celled creature splitting and each copy thinking it was the original. Something like that could be happening in quantum measurement all the time, and, like the amoeba, each version would believe it was unique. Wheeler found the metaphor misleading (experimenters are not amoebas) and axed it.

Despite Wheeler's heavy editing of Everett's work, Bohr was completely uninterested. He and Petersen saw no problem with the existing interpretation of quantum measurement. What we know, they believed, is based on the measurer's particular methods, such as his choice of apparatus, and the results he obtains from doing certain experiments. From that, we might try to infer what the quantum state must have been like to produce such measurements. The idea that we know enough about the quantum state to draw conclusions about its effects on and by the experimenter seemed audacious, unwarranted, and unnecessary.

ACROSS THE WAVES

From January until September 1956, Wheeler took a research sabbatical as a Lorentz Visiting Professor in Leiden, Holland. Of his family members, he brought along only Janette and Alison, as Letitia and Jamie had already started college. Three of his graduate students joined him as well: Misner, Putnam, and another young American, Joseph "Joe" Weber. Weber was interested in exploring the properties of gravitational waves: ripples in the fabric of spacetime emanating from fluctuating massive bodies (such as collapsing stars) like sound waves emitted by the vibrating diaphragms of speakers. Everett, who had not yet defended his thesis, elected not to come along.

Wheeler headed up to Copenhagen in May and met with Bohr and Petersen to try to convince them of the merits of Everett's hypothesis. Wheeler needed the universal wave function concept to eliminate the need for external observers and enable description of the entire universe in quantum language. There was no other reasonable alternative. Clearly, no mortal scientist could hop on a spaceship, travel beyond the universe, observe it, and cause its wave function to collapse, just to snag a measurement.

Wheeler urged Everett to join them in Copenhagen for the discussions. Instead, Everett was getting ready for a nonacademic summer job at the Pentagon. Bohr and Petersen wouldn't budge in their opposition to the concept, prompting Wheeler to urge Everett to further edit his thesis before he received his PhD. In the end, the dissertation was so watered down and nondescript that few who didn't already understand Everett's notion could fathom what he was trying to say. Discouraged, he left academia for a military research career. (In 1959, taking a break from his job, Everett would finally travel to Copenhagen, but once again Bohr wouldn't shift his stance.)

Joe Weber was another dreamer who admired Wheeler's openness to unconventional ideas. His fervor for gravitational waves tugged Wheeler in yet another direction. Einstein had predicted such creases in reality's fabric in one of his early papers on general relativity. He wavered in his support before embracing the concept again in a 1936 paper with Nathan Rosen.

Weber calculated the impact of such waves on Earth and concluded that the effects would be faint. Nevertheless, perhaps a sensitive enough detector could pick up the ripples from a stellar catastrophe, such as a colossal supernova explosion that marked the end of a massive star's life. If the initial burst of energy were great enough, perhaps the ripples might still be felt thousands of trillions of miles away.

The prospect of detecting gravitational waves mesmerized Weber and shaped his career for many decades. At the University of Maryland, he would design a room-sized apparatus, called a "Weber bar," to try to pick up their faint signals. While he would report evidence for such waves, others could not reproduce his results. As it turned out, a much larger and far more sensitive detector would be needed—the Laser Interferometer Gravitational-Wave Observatory (LIGO)—which first achieved success in September 2015 (announced in early 2016), detecting ripples produced by a distant pair of black holes, spiraling toward each other in their death throes. Another of Wheeler's students, Kip Thorne, who became a professor at Caltech, was a founder and leader of that project.

In the mid-1950s, however, when Weber studied under Wheeler, gravitational waves were a controversial topic. At a 1955 celebration in Bern, Switzerland, of special relativity's jubilee, Rosen argued that they didn't carry energy. He based this notion on calculations showing that the gravitational energy must be clustered in the vicinity of stars and other massive objects, not empty space. At a conference held in Chapel Hill two years later, though, Feynman offered a simple line of reasoning for why gravitational waves should have energy, which became known as the "sticky bead argument."

STICKY BEADS

In September 1957, the DeWitts—with Cécile as the driving force—organized the first major international conference devoted to general relativity held in the United States. Sponsored by the Institute of Field

Physics at UNC Chapel Hill and funded by Bahnson, it was so important that it became known simply as "GR1." As Bryce DeWitt recalled, "[It] was a conference by invitation only. The people there were, among others, John Wheeler, Leon Rosenfeld, Tommy Gold, Fred Hoyle, and Richard Feynman. It was a lovely conference."

Wheeler brought along several of his students, including Misner and Weber, who each presented. While Everett didn't attend, Wheeler sent DeWitt a copy of the drastically cut version of his thesis, with the nondescript title "'Relative State' Formulation of Quantum Mechanics," for inclusion in the conference proceedings, along with a second article summarizing Wheeler's own take on it. Also included were eight contributions by Wheeler and his students, making up more than one-third of the total. Other conference attendees gently ribbed him for his "takeover" of the event.

Feynman agreed to attend the conference as a kind of outsider who, while not an expert in gravitation, could tell the participants if their ideas made any sense. When he arrived at the Raleigh-Durham airport on the second day of the conference, he walked up to a taxi stand and asked for a driver to take him to the University of North Carolina. However, because North Carolina State University is in the same region, the dispatcher wasn't sure if he wanted to go there or Chapel Hill. Feynman didn't know either but had an idea. He asked if there had been a lot of requests the previous day from a bunch of distracted people muttering "G-mu-nu" (a term from general relativity). The dispatcher immediately knew exactly what he was talking about and ordered a cabbie to drive him straight to Chapel Hill.

When Feynman arrived at the conference, he decided to tease Wheeler about his latest crazy scheme. As DeWitt recalled, "Feynman showed up and he said, 'Hi Geon.' He called him Geon Wheeler."

Feynman wasn't alone in doubting geons. "Nobody believed in it," said DeWitt. "But Wheeler was trying to do the same as I. He was trying to drag general relativity back into the mainstream of physics. He was taking an engineering approach. Just trying to take this esoteric mathematical thing and make it something you can grab hold of and talk about in a physical way."

Along with geons, Wheeler presented at GR1 a grab bag of oddities related to geometrodynamics, emphasizing the idea of wormholes and the foaminess of spacetime on its tiniest scale. He introduced a clever metaphor about the human experience of spacetime being like the view of an ocean from an airplane, smooth and calm. But if we could get down to ground level, he argued, it might seem turbulent and frothy. Similarly, spacetime at the Planck scale might well be bubbling with ephemeral interconnections such as transient mini-wormholes.

One key point of discussion at the conference was whether empty space might transmit energy in the form of gravitational waves. Participants brought up prior arguments against that possibility, such as Rosen's assertion that they have zero energy. However, after comments by Feynman about the necessity of quantizing gravitation, Leon Rosenfeld pointed out that applying quantum methods to gravity would likely require a comprehensive description of gravitational radiation, analogous to that for electromagnetic waves. Otherwise, physicists wouldn't have a clue where to begin. Therefore, gravitational waves had better exist, or quantum gravity might never get off the ground.

After thinking deeply about the question, Feynman thought of a simple rebuttal to Rosen's negative conclusion. Picture two masses near each other, but not touching, which are free to move—like beads on an abacus. They are connected to the same stick—the first one firmly and the second free to slide along it. Now imagine a gravitational wave passing through that part of space and jiggling the masses. The second mass would rub against the stick, creating heat due to friction in the process. Given that, due to conservation laws, the energy needed to come from somewhere, clearly it was transmitted via the gravitational wave. Ergo, gravitational waves must convey energy.

If Feynman had been alive in 2015, undoubtedly he would have been excited to see his hunch confirmed by LIGO's discovery of gravitational waves. His life did overlap for a few years with the program. Thorne cofounded the project in 1984, along with Rainer Weiss and Ronald Drever, and saw it to success after many decades of planning. Fittingly, Thorne's academic title at the time of the discovery was Feynman Professor of Theoretical Physics, Emeritus at Caltech.

THE MANY WORLDS INTERPRETATION

After the Chapel Hill conference, DeWitt assumed the task of editing a volume of *Reviews of Modern Physics* devoted to its papers and discussions. Because Everett hadn't attended and his work hadn't been discussed, the inclusion of his paper—submitted by Wheeler along with an analysis—was a bit of a mystery. The title seemed well matched with the quantum part of the conference but not so much with its theme of gravitation—except for the general idea of a universal wave function (quantum descriptions without collapse). Nevertheless, as it had Wheeler's backing, DeWitt read it over very carefully. While at first "tickled to death" that someone had brought something novel to quantum measurement, he grew increasingly alarmed by its cavalier references to observers participating in a bifurcation of quantum states. "I was so shocked that I sat down and wrote . . . [a long] letter to Everett, alternatively praising and damning him," recalled DeWitt. "My damning largely consisted of quoting from Heisenberg regarding 'the transition from the possible to the actual' and insisting upon the fact that 'I do not feel myself split.'"

Everett sent a brief reply. In it he took the opportunity to reintroduce, in the form of a footnote, a bit of the explanation of splitting that Wheeler had cut out of the article. He explained how, after a quantum measurement, each copy of the observer thinks that his version of reality is the genuine one. Also, the fact that you don't feel something doesn't mean it is not happening. He asked readers (and DeWitt) to recall the anti-Copernicans in Galileo's day who argued fallaciously that Earth didn't revolve around the sun because no one felt it moving. Marveling at Everett's witty comeback, DeWitt exclaimed, "Touché!"

Everett's hypothesis remained little known until 1970, when DeWitt penned a popular description of it for *Physics Today.* He redubbed it the "many worlds interpretation" (MWI) of quantum mechanics, certainly a far more descriptive name than "relative state." The piece was much discussed, conveying the concept to a much wider audience.

DeWitt also decided, with Everett's permission, to release a scholarly book about the MWI. Everett contributed to the volume by sending him a rumpled copy of an early draft of his thesis, before Wheeler had put it on the chopping block. From that prototype, DeWitt got a much clearer picture of the concept.

For the rest of his career, DeWitt was the MWI's most ardent advocate and popularizer, stressing how any quantum description of the universe could not have external observers. Therefore, the MWI was the only game in town. Nevertheless, he could fully see why others would doubt the idea that reality split ceaselessly into myriad copies. Thanks in part to DeWitt's advocacy and to subsequent enhancements by respected physicists such as David Deutsch, who did postdoctoral research under him and Wheeler, and Max Tegmark, who collaborated with Wheeler, the MWI has become a respected alternative (in some circles at least) to the Copenhagen interpretation.

Wheeler maintained mixed feelings about the MWI. In truth, he saw much to like in the universal wave function but was very uncomfortable with terms such as "many worlds," "parallel universes," and "splitting." Why bring in more than one universe?

Feynman largely ignored the MWI interpretation. He uttered one of his few recorded comments about it at the Chapel Hill meeting, following a talk by Deser. After Wheeler mentioned the possibility that Everett's notion of a universal wave function might be easier to apply to gravitation than the standard techniques of quantum electrodynamics, Feynman responded, "The concept of a 'universal wave function' has serious conceptual difficulties. This is so since this function must contain amplitudes for all possible worlds depending on all quantum-mechanical possibilities in the past and thus one is forced to believe in the equal reality of an infinity of possible worlds."

As Dyson later noted, "Feynman had no use for philosophy and disliked equally all philosophical interpretations of quantum mechanics. He said, the theory is clear and simple if it is not obscured by philosophical fog. The purpose of theory is to describe nature, not to explain nature." Dyson similarly thought the MWI of little value. As he remarked, "I do not remember when I first heard of the Everett interpretation. I always disliked it and considered it a stupid waste of time to discuss it. Borrowing the words of [Wolfgang] Pauli, I would say that it was 'not even wrong.'"

Sum over histories and the MWI both postulate parallel strands of reality—veritable labyrinths of time. Yet, while the former has become an accepted means of describing particle physics, the latter remains controversial. One might think that parallel universes are all the same, but the two approaches have key philosophical differences. In sum over histories, in conducting a quantum measurement we experience a blend of different paths through spacetime, enacted by particles in an abstract realm of possibilities. However, the mixture cannot be separated into different physically observable segments. There is always only one universe and one classical reality.

MWI, in contrast, makes the splitting real. The world around us—including our very beings—divides repeatedly in an ever-expanding web of chronologies. As in Jorge Luis Borges's "The Garden of Forking Paths," along one branch two people might be close friends and in another, mortal enemies. Perhaps by some quirk of fate, in some zany parallel universe, Feynman would be acting in the film *Some Like It Hot* with Marilyn Monroe, pounding on conga drums in the various jazz band scenes, while professors Jack Lemmon and Tony Curtis presented a scholarly paper at Chapel Hill. Feynman surely would have loved that.

Such alternative realities seem too "science-fictiony" for many hard-nosed physicists. Even for someone like Wheeler, who relished "crazy ideas," the notion of conduits to actual parallel universes was a bridge too far. For him, untestable assertions bordered on religious credos rather than representing authentic physics. Dream at night, but verify in the morning light.

CHAPTER SEVEN

TIME'S ARROW AND THE MYSTERIOUS MR. X

We drop an egg on the sidewalk: it splatters in all directions. On the other hand if we had a smear of egg on the sidewalk, we would not expect it to come together to form a complete egg and ride back into our hand. So it would be obvious that the laws of nature appear different if we were to reverse the direction of time.

—Richard P. Feynman, from notes for the "About Time" program, reprinted in *The Quotable Feynman*

Time. Why is there any such thing as time? Why should it be one-dimensional? Time cannot be fundamental. The ideas of "before" and "after" fail at very small distances. They fail at the big bang.

—John A. Wheeler, quoted in Jeremy Bernstein, "What Happens at the End of Things?"

There is a marked asymmetry between our memories and our aspirations, our reflections on the past and our hopes for the future. Perhaps that is for the best. If we were able to "remember" the future, we might have difficulty embarking on a relationship or creative project if we knew in advance that it would end in failure.

If he'd had a crystal ball in 1952, Richard Feynman would have realized from the start that his marriage to Mary Lou was doomed. He might not have pursued superconductivity if he knew that others—John Bardeen, Leon Cooper, and J. Robert Schrieffer—would identify the major features of its theoretical description (although he generated important insights about it and made significant contributions to its sister field, superfluidity). Similarly John Wheeler, if he could have peered into the future, would have avoided bringing a classified document aboard a train. He might have paid little heed to geons, knowing in advance that they'd prove unstable and untenable as particles, remaining theoretical artifacts that ultimately vanished into obscurity.

Triumphs, as well as missteps, often surprise us. In 1957, Feynman and Wheeler had no way of knowing that the coming decade would generally prove fruitful and happy for each of them (aside from Wheeler's mother's death in 1960). Feynman would finally achieve a thriving marriage, experience the joy of having children, develop the acclaimed educational series *The Feynman Lectures on Physics*, discover critical features about elementary particles and the forces of nature, and make key suggestions that helped spur the field of nanotechnology—not to mention receiving the Nobel Prize for his earlier contributions to quantum electrodynamics. Wheeler would experience the pleasure of seeing his children get married; the happiness of becoming a grandfather; the honor of receiving the Enrico Fermi Award, the Franklin Prize, and the Albert Einstein Award; and the satisfaction of becoming a respected authority on the gravitational collapse of stars, developing and promoting the notion of black holes.

In contrast to humans, elementary particles, of course, have no way of fathoming either their past or their future. If they somehow could, would they even notice the difference? Up until the early 1960s, most physicists believed that (aside from measuring processes in which humans intervened) all elementary particle interactions were completely time reversible. Film a particle interaction, run that video backward, and the reversed version would be just as common and credible as the time-forward picture.

Then a remarkable discovery in 1964 by James Cronin and Val Fitch, working at Princeton, proved that even subatomic bodies might, in some cases, display a distinction between past and future. They showed that time-reversal symmetry is not a universal feature of the particle world. Rather, for some processes, there is a one-way arrow of time.

REFLECTING ABOUT REFLECTION

To appreciate Cronin and Fitch's discovery, we need to understand the concept of symmetry in particle physics. Some symmetries are continuous, such as invariance under rotation. Twirl a hydrogen atom in its ground state (lowest energy level), and its measured physical properties will appear the same. Translational (movement through space) symmetry is also continuous: nudge the same atom a little bit through empty space, and it similarly stays the same.

Other types of symmetry are discrete, involving a shift between a finite set of configurations. Charge conjugation (switching of signs) symmetry, labeled *C*, is one example. If you switch a particle from positive to negative and nothing else changes, *C* symmetry holds. Another example is parity invariance symmetry, labeled *P*. Parity transformation means reflection in a mirror. Mathematically, it involves reversing the sign of one or more spatial coordinates—from positive to negative, or the converse. If *P* symmetry holds, changing the direction of an interaction to its mirror image doesn't affect the result.

Imagine that you work for a recycling center. Someone hands you a paper glove to be turned into pulp. You don't care if it is left-handed or right-handed; it is all the same. *P* symmetry holds. In contrast, if in a baseball game you are a right-handed catcher issued a left-handed mitt, you will almost certainly exchange it. In that case, *P* symmetry does not hold.

The difference between left-handedness and right-handedness is called "chirality." Many familiar things—gloves, shoes, seashells, doors, and so forth—have a certain handedness. In *P* symmetry, changing the chirality does not affect the outcome.

Time-reversal symmetry, or *T*, is yet another discrete symmetry, one with two choices: forward or backward in time. Hitting two pool balls together to create a perfectly elastic collision is an excellent example of a case where *T* symmetry seems to hold. The growth of a baby into an adult is one of zillions of counterexamples in the natural world. Clearly, a video of human development looks markedly different played forward or backward in time—a case of *T* symmetry violation on the mundane scale.

Finally, there are near-symmetries, in which a switch causes a slight difference. Protons and neutrons have almost the same mass, obeying a near-symmetry called "isospin." Near-symmetries sometimes inform us about connections between particles. In the case of protons and neutrons, it turns out that both are comprised of subconstituents called "quarks" and "gluons."

Feynman diagrams, with perpendicular axes depicting space and time, along with lines and squiggles tracing typical particle paths, offer an outstanding means of exploring various symmetries. One might use them to show how a combination of *C*, *P*, and *T* symmetries must remain invariant for all known particle interactions. For instance, take an electron moving rightward. Switch its charge, and it transforms into a positron moving rightward. Reflect it in the mirror, and it becomes a positron moving leftward. Now reverse the direction of time, and it turns into a positron moving leftward while traveling backward in time. Based on Feynman and Wheeler's concept of positrons, we represent this by switching the directional arrow in the diagram, resulting in an electron traveling rightward while moving forward in time. We have gone full circle, showing that the magic combination of *CPT* equates to no change at all.

Another way of saying this is that performing any two of those transformations combined is the same as doing the third. The combination of *CP*, for instance, is equivalent to *T*. If *CP* is invariant, so is *T*. Violate *CP*, and *T* is violated as well.

For electromagnetic interactions, each of the three transformations is invariant. You can observe that by measuring the repulsive force between two electrons, one on the left and the other on the right, and applying any of the transformations. Applying *C*, switch both charges to positive, and the force is the same. Applying *P*, swap left and right, and the force is the same. Now run the interaction backward in time, *T*, and the force is the same. Comforting, but boring.

More interesting is the weak interaction, involved in many forms of radioactive decay. Early models of that force assumed all three invariances as well. However, in 1956 physicists C. N. Yang and T. D. Lee postulated that certain types of kaon (K meson) decay demonstrated a violation of parity symmetry. The weak interaction, they argued, possessed a bias between certain processes and their mirror images—like the imbalance between the percentage of right- and left-handed people. Experimentalist C. S. Wu splendidly confirmed their hypothesis, resulting in a Nobel Prize the following year for Yang and Lee.

A LEFT-HANDED PITCH

Feynman rarely collaborated, as he enjoyed the fun of working things out on his own. Also, any collaborator would need to respect his changing moods. While generally optimistic and energetic, on some days he simply wanted to be left alone. Walk into his office on such an occasion, and he might brusquely ask you to leave. He was also blunt in his opinions. If he didn't care for an idea, he might call it dumb or stupid; if he wasn't interested in it, he might pretend to doze off. Hand him a paper, and unless it immediately caught his attention (as in the case of Wheeler's geon article), it would usually lie unread. It was better to be perfectly honest, he felt, than to waste precious time beating around the bush.

Nonetheless, one of Feynman's most important contributions was collaborative, in a sense, with one of his Caltech colleagues, Murray Gell-Mann, a brilliant physicist in his own right. Each was working on remaking the weak interaction in a way that accounted for parity violation. Because they developed similar mechanisms and were from the same institution, they decided to combine forces and write a joint paper.

One of Feynman's motivations for trying to tackle a novel field of study was a feeling of unease about his achievements in physics. His theory of quantum electrodynamics's need for renormalization had made it seem mathematically suspect and incomplete. Despite universal acclaim for the astounding predictive power of that theory, he dismissed renormalization as a "scheme for pushing a great problem under the rug." Failure to unlock the key to superconductivity had been frustrating as well.

The weak interaction seemed a promising target. Aside from Enrico Fermi's fledgling efforts, it remained largely unexplored and ripe for major breakthroughs. Feynman had always wanted to move beyond electrodynamics, and the weak interaction offered ideal grounds for investigation. In the summer of 1957, it occurred to him that a combination of vector (V) and axial-vector (A) interactions would produce a parity-violating model that nevertheless conserved other physical quantities such as charge. The difference between a vector and an axial-vector entails a reversal in direction that the latter picks up during reflections.

To see the difference, stand in front of a mirror, smile, hold out your left hand, and make a thumbs-up sign. If your reflection similarly smiles and makes a thumbs-up sign, with what looks like its right hand, then your thumb represents a vector. If your thumb, as seen in the mirror, somehow points down, even though the fingers are curling in the expected way, it represents an axial-vector. Weirdly, it would seem that your mirror reflection's left hand was where its right hand ought to be and was forced to point its thumb down because of the way its fingers were curling.

Feynman found that the combination of the two situations, V-A, led to the astonishing property that neutrinos are always left-handed—meaning that a neutrino's spin (spin up or spin down) always aligns in the direction opposite to its motion. It would be as if all footballs thrown in the air rotated clockwise from the perspective of the direction they were tossed and never counterclockwise. In the case of a football, you can always slow it down and start it spinning in the opposite direction. But according to the information available at the time, neutrinos had to travel at the speed of light because they were thought to be massless (we now know that they have tiny masses). Therefore, they could not be slowed down and flipped.

With neutrinos always being left-handed and interacting via the weak force with other left-handed fermions, the cosmos was unbalanced. No wonder parity was violated: the mirror image of left-handed neutrinos couldn't be right-handed neutrinos, as they didn't exist. Theory matched experimental data in showing that mirror symmetry wasn't fundamental in nature; it applied to electromagnetism but not to the weak force. When Steven Weinberg, Abdus Salam, and Sheldon Glashow each contributed to the Nobel

Prize–winning formulation of a unified theory of the electroweak (electromagnetism combined with weak) interaction, the V-A mechanism was built right in.

Feynman was extremely proud of his V-A formulation. As he told scientific historian Jagdish Mehra, "As I thought about it, as I beheld it in my mind's eye, the goddamn thing was sparkling, it was shining brightly! As I looked at it I felt that it was the first time, and the only time, that I knew a law of nature that no else knew. . . . I thought, 'Now I have completed myself!'"

While reveling in his description of the weak interaction, he didn't realize that others had arrived at the same conclusion. As he came to learn, physicists E. C. George Sudarshan and Robert Marshak of the University of Rochester had already identified the V-A mechanism. Based on work completed a few months earlier, they submitted a paper to a different journal at almost exactly the same time as Gell-Mann and Feynman's submission to *Physical Review*. Nonetheless, Feynman remained gratified that he had contributed to one of the newly devised laws of nature.

LADY OF THE LAKE

In September 1958, Wheeler and a number of other prominent scientists from the United States and Western Europe, as well as the Eastern Bloc (the part of Europe dominated by the Soviet Union), converged at Geneva's United Nations complex for the Second International Conference on the Peaceful Uses of Atomic Energy. (Wheeler had also attended the first international conference, held in 1954, when he was working on geons.) One special aspect of the meeting was that Americans and Soviets compared notes, for the first time, about their closely guarded schemes for nuclear fusion. They shared not the weapons aspect but rather the potential for energy production. Wheeler enjoyed meeting with Soviet scientists in a friendly atmosphere, offering hope that the Cold War would soon peacefully conclude.

Feynman was invited to contribute a plenary talk to the meeting about particle physics, summarizing his and Gell-Mann's ideas about the state of the field. While in Geneva, he decided to visit its beautiful lakefront, where he struck up a conversation with a pretty young woman in a polka-dot bikini relaxing on the beach. Her name was Gweneth Howarth, and she hailed from Yorkshire, England. She was working long hours as an au pair for an English family in Switzerland for a meager wage. Feynman took a liking to her and mentioned the possibility of her becoming his housekeeper in the United States for a higher salary. While she had hoped to travel around the world, including a trip to Australia, she agreed to mull over his offer.

Back in Southern California, Feynman investigated the legal nuances of bringing her over. To avoid potential complications just in case they ended up having a romantic relationship, he found a colleague to sponsor her, enabling her to obtain a work visa. She arrived in June 1959 and stayed in a separate part of Feynman's Altadena home. While she served as his housekeeper, he remained interested in her. The first year their relationship was primarily professional. While they occasionally went out together, both dated other people. By the following year, however, he had realized he was in love with her and proposed marriage. She accepted on the condition that, for her family's sake, they be married by a minister, not a judge.

On September 24, 1960, Dick and Gweneth were wed. Wesley Robb, dean of the University of Southern California's divinity school, performed the ceremony. The best man was Armenian artist Jirayr "Jerry" Zorthian. It was a strong marriage, built to last. Gweneth understood him very well and gave him ample freedom to pursue his dreams. They would have many fun times together, traveling to different places around the world and raising two children: Carl, born in 1962, and Michelle, adopted in 1968.

Zorthian, who had become one of Feynman's dearest friends, was quite a character. They had met at a party. As Feynman was serenading the partygoers with bongo rhythms, Zorthian decided to join in as a wild dancer. He slipped out of the room for a minute, returned with shaving cream decorating his bare chest, and gyrated like a madman to the beat. Feynman was impressed and delighted.

Feynman loved Zorthian's art and wanted to learn from the master. They decided to make a deal, teaching each other painting and science in alternating sessions. Under Zorthian's tutelage, Feynman's natural drawing talent blossomed into a kind of second career as an artist. He would sketch models, draw portraits of female dancers in a local go-go bar, render people he knew, and take every opportunity to practice. Eventually he sold some of his works under the pseudonym "Ofey."

THE GREAT EXPLAINER

At Caltech, Feynman was much beloved by students as a brilliant lecturer. His "apprenticeship" as Wheeler's teaching assistant and his years at Cornell honing his craft had served him well. With boundless energy, he aspired to be students' favorite, most hilarious, mind-blowing professor. Walking into class, he'd often start the drama by beating on a desk, stand up in front of the room, challenge the students with a profound question about nature, and begin his wild antics.

As the *Los Angeles Times* reported, "A lecture by Dr. Feynman is a rare treat indeed. For humor and drama, suspense and interest it often rivals Broadway stage plays. And, above all, it crackles with clarity."

Feynman started a course called "Physics X," aimed at first-year undergraduates. The instructions for class time were simply "Ask me anything." The course would run for many years and become an emblem of his innovative teaching.

In the courtyard of Dabney House, an undergraduate dormitory, was a bas-relief depicting famous scientists and philosophers, such as Euclid, Archimedes, Isaac Newton, and Leonardo Da Vinci, honoring a person standing behind a podium in the center thought to be Galileo. In 1965, students took the opportunity to label the central figure "Feynman," demonstrating how much they loved and respected him. Some of the freshmen called it the "Oracle of Feynman" and bowed to it for help in physics.

Feynman's reputation as the "Great Explainer" would spread nationally and internationally throughout the 1960s, 1970s, and beyond. He'd be invited to give a recorded series of lectures at Cornell in 1964, titled *The Character of Physical Law*, and appear on numerous science television programs in the United States and the United Kingdom (for example, "The Pleasure of Finding Things Out," recorded in 1981, and "Fun to Imagine," recorded in 1983).

He delivered one of his most influential lectures, "There's Plenty of Room at the Bottom," on December 29, 1959, at an American Physical Society meeting hosted by Caltech. In the talk, he spoke about the enormous promise of miniaturization, from encyclopedias printed on the head of a pin to tiny motors. He offered $1,000 prizes for those able to accomplish such feats—a challenge quickly accepted. The following year, Pasadena engineer William MacLellan sent Feynman a working miniature motor measuring one-sixty-fourth of an inch on each side, much smaller than the head of a pin. Feynman congratulated him and promptly mailed him a check.

Feynman's 1959 lecture is often seen as one of the driving forces behind the field of nanotechnology. Indeed, he was prophetic in his predictions about computers and related technology becoming tremendously more compact—shrinking from room-sized processing devices to today's pocket-sized smartphones. He would continue to maintain an interest in computation and technology, a vestige of his role at Los Alamos during the war as one of the lab's trusted computer experts.

THE SECRETS OF MISTER X

In spring 1963, Wheeler and Feynman found themselves attending yet another conference together and learned that they were being celebrated for a theory they had long discarded. Neither had given much thought to the Wheeler-Feynman absorber theory for many years. Wheeler's "everything is fields" philosophy had pushed action at a distance to the wayside. Feynman saw his PhD thesis as the genesis of the sum-over-histories method but flawed in other respects. Much to their surprise, an article published in 1962 by physicist J. E. Hogarth had resurrected their idea and applied it to cosmology as an explanation for the forward arrow of time.

Hogarth reprised Wheeler and Feynman's calculations in which an equal mixture of signals traveling forward and backward in time led to the radiation-damping effect (reduction in the acceleration of a charged particle as it moves through space). However, he placed the system in an expanding cosmology of the Steady State variety instead of in static space. Because space is expanding, it absorbs signals.

Hogarth showed that the direction of the "cosmological arrow" (expansion of the universe) matches the direction in which radiation damping has the correct mathematical sign. His implication was that the Steady State type of expansion constituted the correct cosmological model because it produced the right kind of particle dynamics.

Hogarth's take on the Wheeler-Feynman absorber theory had become a central focus of the "Nature of Time" conference, organized by Steady State cosmologists Hermann Bondi and Thomas Gold and held at Cornell. Fellow Steady Stater Fred Hoyle was in attendance, along with his gifted young graduate student Jayant Narlikar. They delivered a talk right after Hogarth, which even more explicitly connected the arrows of time and the absorber theory to the predictions of the Steady State model. Other prominent attendees included astrophysicist Dennis Sciama (who was, at the time, pro–Steady State), mathematician Roger Penrose, philosopher Adolf Grunbaum, Charles Misner, Philip Morrison, Leon Rosenfeld, and many others.

Feynman was cynical about the conference program, fearing that it would veer into hypotheses that he didn't support. For example, he felt that the thermodynamic arrow of time (increasing disordered energy) was primary and not necessarily related to cosmology. The choice of cosmological model, he believed, should be based on astronomical evidence rather than guided by theoretical arguments pertaining to time's direction. There was no reason to think that the way space behaved had anything to do with the laws of thermodynamics.

When he found out that the discussions would be recorded, he was very upset. He spoke with Gold, and they reached a compromise. Everything he said would be attributed anonymously to "Mr. X." Nothing would indicate that he was in attendance. Readers would be requested not to quote what he said—as recorded in the proceedings—given that his comments could not be verified.

As Narlikar recalled, "Feynman was quite explicit on this. He felt that knowing that the proceedings are being recorded would make him temper his remarks whereas he wanted to be uninhibited. Mr. X was the compromise arrived at. It is a nice exercise for the reader to solve for X, assuming of course that he was not a participant."

Sheltered by anonymity, Feynman felt free to speak his mind. As was typical for him, he could get quite impassioned if he disagreed with a speaker. Narlikar remembered one such incident:

> I recall the talk given by Dennis Sciama in which he wrote the solution of the wave equation as a volume plus surface integral. For some reason Feynman did not like it. After some argument he agreed to the step with the caveat that the surface integral should not be taken to infinity. He probably felt that limit to infinity may not be well defined. Allowed to proceed, after a couple of steps, Dennis said: "Now I take the surface integral to infinity." And Feynman started rising from his seat, threatening to come to blows, but was held back by Philip Morrison and Tommy Gold!

Feynman's moniker, "Mr. X," served as a kind of inside joke with himself. It was consistent with his desire to maintain a reputation as an average guy with proletarian tastes who happened to possess the uncanny ability to crack the most challenging puzzles in record time. He didn't care if other people were impressed by his title or his status; those were superficial and meaningless. Rather, he wanted to delight them with his humor, amuse them with his lack of convention, and wow them with his clever feats.

His course "Physics X" stemmed similarly from his desire to be the "Super-Everyman"—a Clark Kent prepared at any moment to display his wondrous powers.

THE QUANTUM FLUCTUATION AT THE BEGINNING OF TIME

Wheeler's contribution to the "Nature of Time" conference, "Three-Dimensional Geometry as a Carrier of Information About Time," reflected his continued belief that the universe had begun in a chaotic state: a foam of quantum fluctuations in a sea of energy fields and geometry. Within that froth, time would be undefined. The past, present, and future simply wouldn't exist.

Despite such early randomness, the emergence of biological systems on Earth, billions of years later, required a certain amount of orderly energy, a situation physicists refer to as "low entropy." As entropy measures lack of uniqueness, low entropy corresponds to a high amount of order. Such would be needed to kick-start the machinery of time, which churns in the direction of increased entropy.

How then, Wheeler wondered, did the low-entropy situation come about?

He proposed that the universe began with a massive quantum fluctuation that transformed it from utter chaos into a highly unlikely state of low entropy. Ultimately, over billions of years, that reservoir of orderly energy permitted the evolution of life and the existence of conscious observers. Such intelligent beings could observe the universe and surmise what its early conditions must have been like. Hence, in a version of the "anthropic principle," our mere existence places boundaries on the conditions of the primordial cosmos, guaranteeing that it must have enacted an extraordinarily rare kind of fluctuation.

Rebutting Wheeler's concept of a colossal, entropy-lowering fluctuation, "Mr. X" pointed out that it was unscientific to speculate about something so improbable without a shred of evidence. Rather than considering such a highly unlikely scenario, he proposed a different explanation. As the universe progresses over time, more and more of the past becomes known. That growing knowledge represents a measure of order that matches low entropy. In contrast, the future remains unknown; it is comparatively disorganized and therefore has higher entropy. The discrepancy between the low-entropy past (of burgeoning information) and the high-entropy future (of events unknown) results in a natural arrow of time.

Feynman's abstract ruminations about the arrow of time were unusual for him. Generally he preferred not to engage in pure speculation, which he left to the Wheelers of the world. Yet, given the tenor of the conference and his masking behind a pseudonym, perhaps he felt freer to ruminate.

Wheeler continued with his cosmological model. To map out the evolution of the universe over time, he sliced four-dimensional spacetime, like a loaf of bread, into three-dimensional pieces. How each connected with its neighbors, he pointed out, led to a natural definition of time. This reworking of general relativity—from a frozen into a dynamic theory—drew heavily upon the Arnowitt, Deser, and Misner (ADM) formalism, published in 1962.

With general relativity's spacetime sliced up into three-dimensional spaces, Wheeler wanted to apply quantum theory by switching deterministic variables to probabilistic measures. He aimed for the single classical evolution to become a landscape of possibilities. Aspiring to borrow the least action method so brilliantly applied by Feynman to electrodynamics, he hoped to distill the classical path from a quantum jumble of deviant geometries. It was a daunting task. Once again, he would repeat the pattern of developing a classical model, trying to quantize it, and calling on others for help. He would start bugging his friends for advice, including Bryce DeWitt. They'd meet up briefly at an airport—an encounter that would lead to the well-known Wheeler-DeWitt equation, an attempted sum-over-histories approach to quantum gravity. DeWitt recalled,

> I got a telephone call from [Wheeler] one day around 1964 saying he would be passing through the Raleigh-Durham airport, between planes for two hours. Would I please come out there and we would discuss physics? I knew that he was bugging everybody with the question, 'what is the domain space for quantum gravity?' And I guess he had it finally figured in his mind that it was the space of three-geometries. This was not the direction I was really concentrating my efforts, but it was an interesting problem. So I said OK and I wrote down this equation. I just found a piece of paper out there in the airport. Wheeler got very excited about this.

DeWitt knew there would be no external observers to measure the wave function he described. Therefore, he argued, the Everett interpretation provided the only way to deal with it consistently.

Wheeler had anticipated such concerns in the discussions following his "Three-Dimensional Geometry" talk. "The universe is not a system that we can observe from outside; the observer is part of what he observes," he noted. "Everett's so-called 'relative state formulation' of quantum mechanics does provide one self-consistent way of describing such situations."

WHEELER WITHOUT WHEELER

Hoyle and Narlikar both greatly admired Wheeler and hoped that he would see the merits of their perspective. Yet they knew he had given up action at a distance, adopted a field approach, and supported the idea that the universe changes significantly over time—reflecting the Big Bang rather than the Steady State outlook. His concept of a great primordial fluctuation was anathema to anyone who believed in a universe with essentially the same overall composition for all time. As Narlikar noted, "He started by being against field theory but got converted to it and felt very much attached to it. So in spite of our work, which he admired, he would continue in favour of field theory. Since then I have been referring to the Wheeler-Feynman theory as 'Wheeler without Wheeler.'"

The conception of the genesis and early conditions of the universe would take on new importance when, about a year after the conference, astronomers Arno Penzias and Robert Wilson (not Feynman's former colleague but a different scientist by that name) accidentally discovered the cosmic microwave background (CMB) radiation that fills all space—a cooled-down relic of the hot Big Bang. They were searching for galactic radio signals using a horn antenna at Bell Labs in Holmdel, New Jersey, when they noticed a persistent radio hiss. After ruling out sources of ambient noise, they thought it might be due to pigeon droppings on the instrument. However, even after a cleanup, they detected the same noise. They went to Princeton and sought advice from astrophysicist Robert "Bob" Dicke.

As it turned out, Dicke had been looking for background radiation leftover from the Big Bang. Cooled down by billions of years of cosmic expansion, he predicted, it would have a temperature a few degrees above absolute zero. He was thereby astounded when Penzias and Wilson brought in signal data matching his prediction. He was further surprised on subsequently finding out that George Gamow's group had performed a similar calculation back in the late 1940s.

The CMB discovery proved a watershed for acceptance of the Big Bang proposition. Until its data and analysis became known, the scientific community saw Steady State as another realistic possibility. Popular treatment of the rival notions had similarly been fairly balanced. For example, as late as June 1964, the *New York Times* offered the startling headline "Scientist Revises Einstein's Theory" to describe a proposal by Hoyle and Narlikar. The article went on to explain, "The new theory draws on the mathematical reasoning of two Americans who sought to escape from the idea that an electrically charged particle, like an electron, produces an electric field. The two were Dr. Richard P. Feynman, now at the California Institute of Technology, and Dr. John A. Wheeler of Princeton University." The piece quoted Wheeler, however, as pointing out that the Wheeler-Feynman absorber theory broke down under attempts to apply it to the quantum behavior of particles. That was why he no longer endorsed it, despite its making the news nearly a quarter century after he and Feynman had started working on it. Indeed it was "Wheeler Without Wheeler," or more specifically, an earlier incarnation of his evolving thoughts on the nature of reality.

Soon after publication of the CMB evidence, the mainstream cosmology community migrated almost completely into the Big Bang camp. The Big Bang became sanctified as the mainstream theory rather than just a major contender. Some Steady Staters eventually developed an alternative involving smaller "bangs" scattered throughout space such that they would produce a radiation background similar to that attributed to the Big Bang. The key difference is that rather than a unique genesis at a single moment, this revised approach, dubbed "quasi-steady-state cosmology," posits numerous creation events sprinkled throughout the cosmos and transpiring indefinitely backward and forward in time. Today, Narlikar remains a leading advocate of that minority view.

THE LONELIEST PLACE IN THE UNIVERSE

Misner's talk at the "Nature of Time" conference, "Infinite Red-Shifts in General Relativity," began with a bit of humor: "I would like to talk about how people get out of touch with each other," he explained, referring to limits to contact, known as "horizons." "Two observers who were able to talk to each other head off in different directions and eventually there is no longer a possibility of communication between them. . . . This situation occurs in the Oppenheimer-Snyder problem of continued stellar collapse."

"Oppenheimer-Snyder problem of continued stellar collapse" is quite a mouthful. We now call such situations "black holes." But Misner had yet to hear that term.

It is interesting to note how that expression came into general usage. As science writer Marcia Bartusiak explains, around the early

1960s Dicke started comparing gravitationally collapsed objects to the "Black Hole of Calcutta," an infamously crowded eighteenth-century prison. Astronomer Hong-Yee Chiu, who heard Dicke's comparison, may have used the term "black hole" at a January 1964 meeting of the American Association for the Advancement of Science in Cleveland (which Misner also attended). Several magazines covering the event repeated the phrase.

The term would take off, however, in 1967, when Wheeler, hearing the expression from an audience member at a talk, started to advocate its use as a succinct way of describing what he had previously called a "gravitationally completely collapsed object." Subsequently he became known as the coiner of the term—when, even by his own account, he was just its promoter. Today the expression "black hole" ubiquitously describes the cloaked remnant of the catastrophic collapse of a sufficiently massive star at the end of its lifetime.

Having read and digested Robert Oppenheimer and Hartland Snyder's paper early on, Wheeler had originally doubted the whole concept. He knew that it rested on the Karl Schwarzschild model, among the first, and simplest, solutions of Albert Einstein's general relativity equations. There were several issues, though, with applying that model. For one thing, a threshold, now called the "event horizon," delineated where space and time "switch places" by exchanging signs (from positive to negative for space and the converse for time). How could anything pass through such a bizarre frontier? Also, if a mammoth star's dynamics were complex enough, what guaranteed that its core would collapse into a compact object modeled by such a simple solution as Schwarzschild's, defined by only mass and radius?

Mathematician Martin Kruskal, whom Wheeler knew from Project Matterhorn, helped elucidate the nature of a black hole's event horizon by recasting the Schwarzschild solution within a new set of coordinates. In the revised coordinate system, the event horizon is not a barrier at all; rather it is a permeable membrane through which anything might pass (going inward, at least). Kruskal told Wheeler about his findings privately. Wheeler was so impressed that he wrote them up and submitted the paper to the journal *Physical Review* under Kruskal's name, without telling him in advance. Initially stunned when he received the galley proofs, Kruskal ultimately allowed publication of the paper in his name.

Subsequent work by Misner and physicists David Finklestein and David Beckedorff, a student working with Misner, showed that the event horizons were one-way portals. Anything could get in, but nothing could escape, not even light. That finding was the basis of Misner's "Infinite Red-Shifts" talk and became one reason why "black hole" caught on as a descriptive term.

Wheeler verified these results for himself and paid close attention to computational models of the collapse of heavy stars, which seemed to point to black holes as the end state for those of sufficient mass. He also noted Roy Kerr's derivation of a black hole solution in 1963, which took into account rotation as well as mass, and Ezra "Ted" Newman's subsequent development of a complete model of black holes that included three parameters: mass, charge, and rotation. Additionally, he paid close attention to Penrose's 1965 proof that the end result of catastrophic gravitational collapse was, in certain cases, a spacetime singularity—a central point of infinite density. Weighing all the evidence, Wheeler went from black hole skeptic to black hole believer.

THE TROUBLE WITH KAONS

The "Nature of Time" conference drew a stark contrast between time's reversibility on the particle scale and its irreversibility on the human and cosmic scales. That dichotomy, for example, motivated Hogarth to revive the Wheeler-Feynman absorber theory, with its balance of signals going forward and backward in time, and show how it could lead to a cosmological arrow of time. Although aware of how parity is violated during certain particle processes, conference attendees were reassured by the apparent invariance of *CP* (charge-parity) and *CPT* (charge-parity-time). Combined, those invariances implied that *T* (time reversal) must hold firm as well.

Seemingly sacrosanct ideas can be overturned very quickly in the world of particle physics. Cronin and Fitch's 1964 experiment showing how *CP* is violated in certain weak processes arrived like a bolt out of the blue. It showed how, at even the tiniest levels, some of time's roads have one-way arrows.

Their experiment involved recording how neutral kaons (K mesons) decay. While under most circumstances such particles decayed into three pions, in rare cases they decayed into just two. This happened in only about 1 in 1,000 decays but nevertheless showed that a decay process once thought forbidden was actually allowed. Such a discrepancy would be impossible if *CP* invariance strictly held, because the twin processes were the same when charges were switched but not when flipped in the mirror. *CP* invariance means that if one of the symmetries is violated, *C* or *P*, both must be violated to ensure preservation of the combination of the two. On the other hand, any breach of *CP*, no matter how slight, means that time-reversal symmetry is no longer absolute.

On the plus side, a broken symmetry might explain other imbalances. For example, today there is an enormous abundance of matter over antimatter. All the stars and galaxies we observe are made almost completely of ordinary matter. We see antimatter in nature only rarely, as a tiny component of the cosmic radiation raining down on Earth. What caused such a colossal discrepancy? Many researchers believe that *CP* violation in the early universe is the leading culprit behind the disappearance of most of its antimatter.

In the Big Bang's fiery cauldron, matter and antimatter must have been produced in equal quantities. Because the universe was so hot and dense, particles and antiparticles continually annihilated each other, forming photons and other massless exchange particles, which in turn transformed back into particle-antiparticle pairs in a great cosmic equilibrium. As the universe cooled down, however, the electroweak interaction underwent a symmetry breaking in which the exchange particles corresponding to the weak interaction (called W^+, W^-, and Z^0) acquired mass, while the photons conveying the electromagnetic interaction remained massless. The heftiness of the weak exchange particles meant that the force was short range. In addition, because they didn't always respect charge-parity symmetry, nature started to become slightly imbalanced. The *CP* violation led over the eons to an ever-increasing bias toward matter over antimatter—resulting in the enormous discrepancy today.

THE RELUCTANT NOBELIST

Major experimental findings, such as Penzias and Wilson's detection of the CMB and Cronin and Fitch's discovery of *CP* violation in the decay of neutral kaons, often attract the attention of the Nobel Prize selection committee in Stockholm. New theoretical methods and insights are sometimes not such obvious picks. In the case of quantum electrodynamics, in 1955 Willis Lamb and Polykarp Kusch shared a Nobel Prize for some of the experimental evidence that engendered the field—namely, the Lamb shift and the anomalous magnetic moment of the electron.

A decade later, it had become clear how valuable Feynman's diagrams, sum over histories, and other techniques had become to particle physics, how significant were the renormalization methods of Julian Schwinger and Sin-Itiro Tomonaga, and how important Freeman Dyson's work had been in bringing the three approaches under the same umbrella. Each year the Nobel selection committee could award the prize to up to three individuals or organizations. Unfortunately that meant, in effect, that Dyson wouldn't be included with the others. In 1965, the Physics Prize was split three ways and awarded to Tomonaga, Schwinger, and Feynman "for their fundamental work in quantum electrodynamics, with deep-ploughing consequences for the physics of elementary particles."

Feynman had some inkling that he might eventually receive the prize. However, when he got a series of phone calls in the middle of the night from various reporters congratulating him and asking for his reaction, he became annoyed and agitated. He had done all that work for fun, not for acclaim.

By that point, Feynman was leading a happy, balanced life that he didn't want disturbed, enjoying a wide range of hobbies such as drumming and drawing. For the latter, schooled by Jerry Zorthian, he was often searching for models, usually female, willing to be sketched in various poses. Gweneth trusted her husband and found no reason to be jealous.

One of Feynman's models was a brilliant young astrophysics graduate student, Virginia Trimble, who had been featured as a teenager in the October 1962 issue of *Life* magazine as an emblem of brains with beauty. One of the first women admitted to Caltech, she was studying the properties of stars and nebulae (gaseous remains of exploded stars) under astrophysicist Guido Münch when Feynman invited her to be his paid subject. As Trimble recalled,

> Feynman spotted me crossing Caltech campus one day, encountered Guido Münch (near Robinson, the old astronomy building), and said something like "I'm hunting. Perhaps you know the quarry." And so, for some months, typically something like alternate Tuesdays, I went to the Feynman residence for a couple of hours, receiving $5.50 per hour (a lot in those days!) and all the physics I could swallow. Gweneth used to bring us orange juice and cookies mid-session.

Trimble would become an accomplished astrophysics professor at the University of California, Irvine, and marry University of Maryland professor (and Wheeler's former student) Joe Weber. Years later, they'd go together to a Caltech exhibition of Feynman's art and encounter a sketch of her.

"Joe took a critical look at one of the back views of an unclad torso," Trimble recalled, "and said 'I've seen that back before.'"

Trimble remembered how the frantic aftermath of the Nobel Prize announcement disrupted Feynman's plans to sketch her that evening:

> Feynman came to my office around eight a.m. that day to cancel our evening drawing session. Luckily I already knew because my mother had been listening to the radio and phoned me around six a.m. We were all always morning people. Feynman was not, but had managed to get into a jacket and tie that morning. When graduate students asked him for a special talk just for them, the subject he chose to address was the absorber theory of radiation, part of his thesis work with Wheeler.

The degree of planning needed for the prize ceremony soon overwhelmed Feynman. He received inquiries from Sweden about putting together guest lists and making other arrangements. The scientific lecture would be doable, but other aspects, such as greeting and thanking the king, sounded too pompous for his tastes. He started worrying that he would screw up, much as he had made gaffes at the formal teas during his Princeton days. The prize was more hassle than it was worth, he thought. He wondered if maybe he could turn it down.

He really couldn't, of course. He and Gweneth flew to Stockholm for the ceremony, where he donned a specially made tuxedo and she wore lovely dresses. During the scientific presentation, he made sure to acknowledge Wheeler's contributions. Perhaps his favorite part was the dancing, where he and Gweneth could let loose.

Afterward, he flew to Geneva, where Victor Weisskopf, who had become director of CERN (the European Organization for Nuclear Research), had invited him to deliver a lecture. He decided that, as a Nobel laureate, he should probably start wearing a suit and tie for his talks. But when he got up on stage dressed formally and explained why, the audience started shouting, "No, no, no!" Responding to popular demand, Weisskopf came up to Feynman and pulled his jacket off. Feynman took off his tie, and he was back down to shirtsleeves—his usual mode of dress. He thanked Weisskopf for dismantling his pretentions.

Being a Nobelist was in some ways a headache for Feynman. Deluged with invitations for speaking engagements, he turned most of them down. The few exceptions involved education, such as speaking at schools or delivering talks about the fun of physics for the general public (some of which were televised, becoming popular programs on the BBC and elsewhere).

Offered many honorary degrees, Feynman invariably said no. He recalled his hard work at Princeton obtaining his PhD and didn't want to water down the meaning of a degree by getting one without earning it.

YOU SAY QUARKS, I SAY PARTONS

After returning to Caltech a reluctant but deserving Nobelist, Feynman was ready for something different. Of the four natural forces, he had already contributed to quantum theories of electromagnetism and the weak interaction and made a valiant attempt at unraveling the mysteries of gravitation. Next on his checklist was the strong nuclear interaction, the force that counteracts electrostatic repulsion to bind protons and neutrons together within atomic nuclei. Understanding of the force had come a long way since the days of Hideki Yukawa, with discovery of many more particles that responded to it. Those that reacted were classified as "hadrons" (from the ancient Greek for "stout"); those that didn't were called "leptons" (from the Greek for "delicate"). Hadrons were subdivided according to their spin properties into baryons (half-integer spin, including protons and neutrons) and mesons (integer spin, including pions and kaons). Examples of leptons included neutrinos, electrons, and muons—each of which ignored the strong force completely.

For a progress report on the strong force, Feynman could look within his own department, where Gell-Mann was a leading innovator. Gell-Mann had won acclaim with two of his ideas: the Eightfold Way and quarks. Gell-Mann proposed the Eightfold Way, named for the eight-step road to enlightenment in Buddhism, as a classification scheme for hadrons that organized them according to several parameters, including charge and a conserved quantum number, called "strangeness," discovered in certain decays. The arrangements revealed certain patterns and symmetries. While some of the groups had eight hadrons—hence the name—others included one, ten, or twenty-seven. The scheme had a gap, which led to predictions of a new particle. In 1964, researchers at Brookhaven National Laboratory detected the predicted particle—called the "omega minus baryon"—completing the scheme and lending important support to Gell-Mann's hypothesis. It was a triumph of the application of symmetry to particle physics.

That year, Gell-Mann demonstrated how his scheme could be explained if baryons were made up of three types of constituents, arranged in different combinations, like cards in a poker hand. He plucked his name for these, "quarks," from one of the most impenetrable books in literature, *Finnegan's Wake* by James Joyce, a stream-of-consciousness novel that contains the line "three quarks for Muster Mark." Gell-Mann liked the sound "quark," which he pronounced like "quart" with a *k*, and noted that any baryon had three of them. One tricky part was that each had fractional charge—either 2/3 or –1/3 the charge of a proton. Antiquarks would have the opposite charges of their quark counterparts. Such fractional charges had never been detected in nature. However, if quarks were always confined, the lack of evidence for fractional charges was not a major issue. Independently, around the same time, physicist George Zweig proposed a similar scheme but called the components "aces."

Feynman was similarly interested in the idea that protons and neutrons had constituents. While certainly aware of Gell-Mann's quark model, he distanced himself from it. When a *New York Times* piece titled "Two Men in Search of the Quark," published in October 1967, implied that the two were collaborating on the topic, Feynman responded with a letter to the editor: "Although I did do many of the things described in your article, I am really not one of the men responsible for starting scientists thinking about quarks. It was the result of one of the great ideas Gell-Mann gets while working separately."

Rather than positing quarks based on symmetry groups, Feynman pursued a phenomenological path, analyzing the results of particle collisions. From that scattering data, he reached virtually the same conclusion as Gell-Mann: that hadrons are made of more fundamental particles. He showed that these must be point particles, like electrons, but responsive to the strong force. Perhaps out of rivalry with his Caltech colleague, he decided to call the constituents "partons" rather than "quarks." "Partons," in Feynman's conception, expressed more of a standard particle feel, while "quarks," in Gell-Mann's original picture, might be more amorphous. Feynman published a paper on the topic of partons in 1969.

While the term "parton" saw some use in the 1970s, the more whimsical expression "quark" won out. We now know that there are six flavors (types) of quarks: up, down, strange, charm, bottom, and top. They vary significantly in mass, with up and down being the lightest and most common. Familiar atomic nuclei are made up of those two alone. The other quark flavors are more exotic, found in energetic cosmic rays and particle debris after high-energy collisions. All the hadrons in nature, or produced in powerful colliders, combine the six quark flavors and their antiquark companions. Baryons are quark trios, and mesons are duets of quarks and antiquarks. For example, a proton is up-up-down, a neutron is up-down-down, and a neutral kaon mixes down-antistrange and strange-antidown.

When theorists developed a quantum field theory of quarks, they looked to quantum electrodynamics and Feynman's diagrammatic methods for guidance. They introduced a new exchange particle, called a "gluon," which would mediate the strong force, as photons do for electromagnetism. A Feynman diagram depicts a gluon as a helix. Oscar Greenberg (who had been in Wheeler's general relativity class when they visited Einstein) thought of a vibrant way of describing the equivalent of electric charge for the strong interaction: color charge. Each quark might have a color charge of red, green, or blue—with a baryon combining all three. These are not real colors; the term is symbolic, in the same manner as "flavor," and has nothing to do with hues. The quantum theory of the strong interaction became known as quantum chromodynamics (QCD).

With the development of QCD, along with the electroweak theory combining quantum electrodynamics with the weak interaction, theorists in the late 1960s and 1970s were very excited about the prospect of grand unification: melding three of the four forces of nature into a unified quantum theory that included quarks, leptons, photons, gluons, and the carriers of the weak interaction. Theorists proposed that at high enough temperatures, such as in the blazing furnace of the Big Bang, all three interactions would have the same strength, range, and other properties. Only when the universe cooled down a bit would those forces start to distinguish themselves.

Eventually the fourth interaction, gravitation, would be stirred into the mix too, theorists hoped. But because of the infinite terms present in attempts to quantize gravity and the notable imbalance in strength between gravity and the other forces, some researchers preferred to focus on the other three first, until those issues could be sorted out. Nonetheless, even attempts to combine the strong and electroweak interactions into a Grand Unified Theory were not completely successful.

Curiously, a stark distinction between the strong and weak interactions had to do with *CP* invariance. The former preserved that combination, while the latter violated it. Given the connection with time-reversal symmetry, it was odd that strong processes appeared the same forward and backward in time, while weak decays could in some cases show a distinction. Couldn't time make up its mind about whether to be reversible, especially if the forces were supposed to be similar at high energies?

THE ALPHA AND THE OMEGA

While particle discoveries, such as Cronin and Fitch's kaon measurements, revealed the weirdness of time on the tiniest scales, cosmological results were similarly confusing. Penzias and Wilson's CMB data demonstrated a remarkable uniformity in temperature, no matter which direction they pointed their detector. That cosmic microwave radiation had been released when atoms were formed, some 380,000 years after the Big Bang. Thermodynamics tells us that temperatures even out only if regions are in thermal contact, meaning that they must be close enough to exchange photons. By then, however, the universe had already had time to develop, and different parts of space were widely separated. Given the lack of opportunity for disparate regions of the cosmos to even out in temperature, why should the relic radiation from that era be so incredibly smooth? That enigma became known as the "horizon problem."

Astronomers knew that the Penzias and Wilson data was not very precise. They realized that superior instruments might reveal minute ripples in the CMB, reflecting slightly denser regions that would form the seeds of structure. Those minor irregularities would grow over the ages, due to gravitational clumping, to form stars and galaxies. Space probes such as the Cosmic Background Explorer, the Wilkinson Microwave Anisotropy Probe, and the Planck Satellite would eventually confirm such hunches, revealing that the CMB is speckled with minuscule fluctuations.

Nevertheless, such subtle variations did not negate the horizon problem, which was temperature uniformity on the largest scale. Wheeler hoped to resolve the issue via a quantum rendition of geometrodynamics. As he had pointed out at the "Nature of Time" conference, perhaps a fully realized theory of quantum gravity could explain why the entropy in the primordial cosmos was so low. Conceivably, sufficiently low entropy would correspond to a uniform early cosmos—much as the low entropy (high level of order) of a deeply frozen pond makes its surface smooth and skateable.

In the meantime, Misner attempted a classical explanation called the "Mixmaster universe." He based his model, proposed in 1969, on an anisotropic solution of Einstein's equations that oscillated in various directions instead of expanding uniformly. His thinking was motivated, in part, by British cosmologist Stephen Hawking's finding that the universe must have begun in a singularity (state of infinite density) and enriched by Russian physicists Vladimir A. Belinski, Isaak M. Khalatnikov, and Evgeny Lifshitz's results showing how space might have emerged from such a singularity in a chaotic fashion. (Wheeler had alerted Misner to the Russians' methods while in the midst of his own calculations.)

Misner dubbed his universe model "Mixmaster" after a kitchen mixer popular at the time. He hoped that its mixing properties—a quirk of the particular cosmological solution—would prove powerful enough to smooth out the universe in its earliest stages, explaining why its radiation was so uniform in temperature. Unfortunately, the blending wasn't robust enough to do so. Mathematically, the solution's churning dynamics didn't mix effectively enough in the time frame required. It left a chunky unevenness, not the milkshake-like smoothness we witness today.

If the beginning of the cosmos presented an enigma, possibilities for its demise offered additional riddles. In that era cosmologists pondered two options for a cosmic finale. The first was a "Big Crunch," a reversal of the Big Bang that would see expansion switch gears and turn into universal contraction. The other was a "Big Whimper," continued but gradually slowing expansion. Such a gradual deceleration would give the stars billions and billions of years to shine, but they would eventually burn out, one by one, leading to the universe's heat death (lack of usable energy). The end, in that case, would be cold and lonely.

Wheeler was strongly interested in the Big Crunch scenario and its repercussions. Along with the Big Bang and black holes, he saw it as a key place to study the extreme conditions of gravity and its effect on time and causality. In an interview, he once called the Big Bang, black holes, and the Big Crunch "the three gates of time."

At a 1966 American Physical Society meeting, he speculated that the Big Crunch would present a rather strange situation when the cosmos started to contract. Suppose cosmic expansion in the Big Bang set time's forward direction and the growth of entropy. Then, he speculated, in the Big Crunch era, entropy might start to decrease, corresponding to a reversal of the direction of time. Biological processes might reverse themselves, resulting in people living their lives backward. Eventually the human race would devolve down to single-cell organisms. As space continued to shrink, Earth would be swept up in a cloud of dust, and the universe would collapse into a point of infinite density. Wheeler conceded, however, that such a time-reversal scenario was highly speculative.

GÖDEL'S CAROUSEL

Reversing the hands of time was an obsession of mathematician Kurt Gödel, a close friend of Einstein when both were permanent members of the Institute for Advanced Study (IAS). Gödel was best known for his incompleteness theorems, published in 1931, showing that no logical system is completely self-consistent. Those ideas helped inspire British mathematician Alan Turing to develop the Turing machine, a systematic approach to computation, which in turn helped motivate John von Neumann's theoretical blueprints for early electronic computers.

In 1949, around the time of Einstein's seventieth birthday, Gödel presented to his friend what he considered a significant discovery: a rotating solution of the equations of general relativity that allowed for backward excursions through time. If the universe had precisely the right amount of spin and the correct type and proportion of matter, enacting certain types of loops through space would enable journeys into the past. Hence, under very special conditions, Einstein's gravitational theory permitted a kind of time travel. While Einstein saw the mathematical validity of Gödel's result, he didn't find it physically relevant. There were so many solutions to his equations; why focus on such an odd one? The mainstream physics community virtually forgot Gödel's rotating model.

By the late 1960s, Gödel was suffering from paranoid delusions that affected his health. Convinced that someone was trying to poison him, he ate only sparingly and lost a lot of weight. He became so emaciated that he needed to wear coats just to keep up his body heat, even on warm spring days. At the same time, he was quietly trying to collect enough proof of an imbalance in the rotational directions of galaxies that would reveal a net spin of the universe itself and hence the possibility of backward-in-time voyages. The idea of undoing time's ravages captivated him.

Wheeler had a strong interest in Gödel's work. Around 1970, when he, Misner, and Kip Thorne were preparing the influential textbook *Gravitation*, they reached out to him for his thoughts. By then, Misner and Thorne had faculty positions elsewhere—at the University of Maryland and Caltech, respectively—so they needed working space to consult with each other about the book. During one of their

Princeton visits to meet with Wheeler, the IAS generously lent them an office in the same building as Gödel's. Out of curiosity, the three decided to knock on his door. When they mentioned their book project, he asked what it would report about rotating universes. Their answer—nothing at all—disappointed him greatly.

Although Wheeler later learned about some data to support an imbalance in the rotation of galaxies and gladly reported it to Gödel, he was not focused on such an idiosyncratic universe model. Rather his emphasis was standard cosmologies and the features they shared with extreme gravitational situations in astrophysics, such as the distorted spacetime environments of black holes.

BLACK HOLES HAVE NO HAIR

The end stages of the Big Crunch bore some resemblance to the ultracompact conditions of black holes. Each involved collapse into a singularity. If time reversed in the Big Crunch, one might ask, what would happen to its direction when one approached a black hole?

In treatments of the region near a black hole using Kruskal coordinates, nothing suggested that time would reverse itself for someone entering. In fact, calculations showed that for an astronaut passing through an event horizon (marking the black hole's point of no return), the clocks on his spaceship would continue to tick forward. Only from the outside world's vantage point would time seem frozen for the hapless explorer.

Sadly, for a Schwarzschild-type black hole, the voyager would be propelled ever closer to the black hole's central singularity. Gravitational tidal forces would stretch him in the direction of his motion, while compressing him in all other directions—a process nicknamed "spaghettification." In short order he would be torn apart. (For a Kerr or Kerr-Newman black hole, the central singularity would be a ring, which in principle might be avoided.)

Outside the event horizon, even with the most powerful telescopes, we'd be spared from seeing any of this mayhem. As Wheeler proposed in 1970 with his student Remo Ruffini, we would observe only the black hole's total mass, charge, and spin (angular momentum, in particular).

To convey the concept that black holes are featureless except for those three parameters, Wheeler coined the expression "Black holes have no hair," which became known as the "no hair theorem." He meant that they were like marine recruits with shaved heads, appearing identical from above. Nevertheless, Feynman took the opportunity to rib Wheeler, accusing him playfully of raunchiness. How could he say such an obscene thing, Feynman jested. Talk about the rascal calling out the rector!

If the universe ended in a Big Whimper, eventually those stars with around the mass of the sun would swell into red giants. Their outer envelopes would evaporate into space, leaving behind white dwarfs. More massive stars would become supergiants and suddenly explode in supernova bursts, leaving behind shrinking, compact cores. Depending on their mass, such remnants would become either neutron stars (made of ultradense nuclear material) or black holes.

The entropy of the universe would either level off or increase, until the universe reached the quiescent state of heat death. However, in his studies of black holes, Wheeler wondered if there might be exceptions. If a black hole gobbled up an object, would it swallow the object's entropy too? Could a black hole thereby circumvent the second law of thermodynamics by reducing the total entropy in the cosmos? Soon, a novel way of defining black hole entropy would provide the answer.

CHAPTER EIGHT

MINDS, MACHINES, AND THE COSMOS

> It's worth a lot of miles to talk with you about anything and everything.
>
> —John A. Wheeler to Richard P. Feynman, November 28, 1978 (Caltech Archives)

> His ideas are strange; I don't believe them at all. But it is surprising how often we realize later that he was right.
>
> —Richard P. Feynman, referring to John A. Wheeler, in "Inside the Mind of John Wheeler", 1986

In the early 1970s, with the introduction of monitors and screens, computer simulations became far more vivid. For the first time, the visual results of calculations could appear in real time, right before your eyes. This was not only useful but in some cases riveting to watch. Coincidentally, that was an era of mind expansion—stimulated by progressive rock; psychedelic art; Eastern philosophy; and, for some, the ingesting of various mind-altering substances.

One of the grooviest experiences for a young programmer or engineer of that era was playing mathematician John Conway's "Game of Life," popularized by Martin Gardner in *Scientific American*. The game simulated biological creatures on a two-dimensional grid populated by zeros and ones, representing the dead (or nonexistent) and living, respectively. It fell into the category of what John von Neumann and Stanislaw Ulam dubbed "cellular automata," simple algorithms that determined how the contents of squares on a lattice were updated systematically according to the values of their neighbors.

The rules for the "Game of Life" determine what happens to zeros and ones (represented on a screen by empty and dark squares) on each iteration. For example, any zero surrounded by exactly three ones "gives birth" to a one. Any one surrounded by more than three ones "dies" from overpopulation and becomes a zero. If its neighbor has two or fewer ones, it dies from underpopulation. Otherwise it survives and remains a one. Based on the rules, patterns evolve from step to step into other configurations, appearing a bit like creatures crawling across the screen, devouring each other, procreating, and otherwise engaged in very lifelike behavior. Aficionados of the game found it enthralling and could play for hours, seeding the screen with various initial configurations, turning on the dynamics, and watching the strangely animate artificial beings explore their world.

Identifying with rebellious youth, Richard Feynman started growing his thick brown hair (speckled with gray) a bit longer and dressing at times even more casually. By that point he was a family man—certainly not a hippie—but still loved the idea of an unfettered life on the road. For a man in his fifties and at an advanced stage in his career, he was incredibly exuberant and open to new experiences. He still didn't have a stodgy bone in his body. The Caltech students treated him like a rock star.

Given Feynman's background in calculations and his interest in nanotechnology, expressed in his 1959 talk, he naturally became increasingly fascinated by the possibilities of computer simulation. Part of his motivation, as the decade wore on, was the strong interest in that field expressed by his son, Carl, who would eventually pursue computer engineering at MIT.

Simulations such as the "Game of Life" suggested that the universe might operate, on a fundamental level, like an automaton processing binary values. A prime advocate of that viewpoint was MIT computer science professor Edward Fredkin, who spent a year as a visiting researcher at Caltech. While predictably cautious about an idea lacking experimental evidence, Feynman would become open to at least discussing it with Fredkin and others. A related question was whether human minds operate like digital processors. Marvin Minsky, another MIT computer scientist and a leading pioneer of artificial intelligence, suggested that the human brain is a kind of processing machine. Feynman came to know Minsky well and was comfortable talking with him about such ideas.

Perhaps even more surprisingly, John Wheeler would come to adopt an information-centered viewpoint as well, linked to quantum measurement theory. In the final stage of his research career, he would abandon "everything is fields" for "everything is information." He would call the idea "it from bit." His advocacy and support were so profound that some computer scientists called him the "grandfather of quantum information."

Among the catalysts for Wheeler's transformation were his interactions with a new generation of students far more familiar with computers and their workings. Niels Bohr had died in 1962 (for Wheeler's students, ancient history), so many were open to novel interpretations of quantum measurement. No longer were deviations from the Copenhagen ideology viewed as heretical. Bryce DeWitt's popular article about Hugh Everett III's many worlds hypothesis, for instance, was very well received. Wheeler applauded DeWitt for his popularization of the idea, while politely disagreeing with terminology like "many worlds" and "parallel universes."

The nature of time entered strongly into Wheeler's new approach. Time's arrow is connected to entropy through the second law of thermodynamics, which mandates its nondecrease as time moves forward. Entropy, in turn, is linked to information theory by means of a formula developed by electronic communications pioneer Claude Shannon that defines a particular "information entropy" for each string of data. Therefore, gaining an understanding of the flow of information is another way of trying to model time. Even in his new emphasis on information and quantum measurement, Wheeler certainly didn't forget about cosmology, gravitation, and black holes. In fact, exploration of the concept that black holes have entropy, and hence store information, was one of his initial forays into information theory.

HIDING THE EVIDENCE

In the early 1970s, Wheeler worked with a number of talented graduate students on questions related to black holes. One of his exceptional students was Jacob Bekenstein, born in Mexico to Polish-Jewish immigrants. The two often discussed the properties of black holes, including the "no hair theorem."

One day, Wheeler was joking with Bekenstein that whenever he happened to place a hot cup of tea next to an iced glass and let them come to equilibrium, he felt like he was committing the crime of increasing the universe's entropy. That is because, according to the second law of thermodynamics, the usable energy due to the temperature difference would become unusable. Ordinarily such a process could not be reversed—slightly hastening the ultimate arrival of heat death. If only he had a black hole to toss the cups into, Wheeler jested. That would conceal the evidence of his crime by forever hiding away any entropy gained. His humorous remark prodded Bekenstein to think about what happens to the entropy of bodies gobbled by black holes.

Inspired by a 1970 paper by one of Wheeler's PhD students, future mathematician Demetrios Christodoulou, showing that during ingestion of materials the surface areas of black hole event horizons always increase or stay the same but never decrease, Bekenstein thought of a brilliant scheme for defining black hole entropy. What if the surface area of the black hole *was* the astronomical way of encoding entropy? The units were different, so there would need to be a proportionality factor. Still, equating the two would offer a natural way of extending the second law of thermodynamics to include black holes. Wheeler wouldn't have to worry about breaking the law if he chucked his beverages into gravitationally collapsed objects.

When Stephen Hawking, who studied black hole properties such as conditions for singularities, learned about Bekenstein's proposal, he was dubious at first. If black holes had entropy, they must have temperature as well, meaning they would radiate into empty space. Anything with nonzero temperature, surrounded by an even colder vacuum, must exude heat. Yet everyone knew that, classically, nothing could escape from a black hole, not even radiation. Nevertheless, Hawking was open-minded enough to calculate what would happen in a simple quantum picture. Much to his surprise he determined that a black hole would radiate extremely slowly into the space around it. That trickle of what became known as "Hawking radiation" would gradually lower its temperature until it eventually reached equilibrium with the space around it—a process that could take many billions of years depending on the black hole's size. Hawking announced his results in a stunning talk titled "Black Holes Are White Hot."

The existence of Hawking radiation and black hole entropy spurred an examination of the question of a black hole's information content. "Information" in this context refers to patterns of zeros and ones, called "bits," in accordance with the notions of Shannon. In his influential 1948 paper, "A Mathematical Theory of Communication," Shannon promoted the idea that every piece of information might be expressed as strings (ordered sequences) of bits, which might be transmitted from one place to another and decoded. That became the basis of today's digital age.

Shannon also defined a kind of entropy—now called "information entropy" or "Shannon entropy"—related to the amount of information carried by a string. That depends on the number of possible outcomes and the likelihood of each. It derives from Austrian physicist Ludwig Boltzmann's much earlier definition of entropy in thermodynamics, which assessed how many possible combinations of microstates (arrangements of particles) might lead to the same thermodynamic macrostate (overall condition such as temperature, pressure, and so forth). If tons of combinations give you the same overall result—such as configurations of fast molecules that correspond to a hot gas—the system possesses high entropy. In contrast, if only rare combinations produce a certain general effect, such as water molecule patterns in a snowflake—that matches low entropy. Shannon translated that idea into arrangements of bits rather than molecules.

Consequently, as Bekenstein and others realized, the area of black hole event horizons not only serves as a measure of entropy but also acts as a gauge of information content. As Wheeler noted, if you divide up the horizon into regions the size of the Planck length squared (the quantum scale), then one bit of information (either a zero or a one) might occupy that tiny area. Therefore, the larger the event horizon in area, the greater the length of the string of binary digits it might store. Wheeler would come to see that connection as the quintessential example of "it from bit": modeling the dynamics of the universe on the quantum scale from an alphabet of binary digits.

HE HAS ALWAYS SOUNDED CRAZY

In 1971, while still in the process of writing *Gravitation*, Wheeler visited Caltech. He took the opportunity to have lunch with Feynman and Kip Thorne at the Burger Continental Restaurant, which served Armenian cuisine. It was a pleasant, relaxing place near campus to meet and discuss ideas.

As they dined, Wheeler explained to his former students his vision that the laws of physics were forged in the Big Bang. Other universes might be out there with completely different laws. There must be some reason our universe has the particular laws it has. Perhaps if it didn't, there would be no life and no conscious entities to experience it.

Wheeler's arguments were a variation of the "anthropic principle," the concept that the universe is the way it is because if it were very different, we wouldn't be here. Such abstract reasoning was anathema to Feynman, as it couldn't be proven or disproven. Parallel histories were fine for quantum calculations, because they supplied experimental predictions that one could readily verify experimentally. But no one could assemble an array of universes and see what happened. So why talk about it?

According to Thorne, Feynman turned to him and offered some sage advice about Wheeler: "This guy sounds crazy. What people of your generation don't know is that he has always sounded crazy. But when I was his student I discovered that if you take one of his crazy ideas and you unwrap the layers of craziness from it one after another like lifting the layers off an onion, at the heart of the idea you will often find a powerful kernel of truth." Feynman then recounted to Thorne how Wheeler's "crazy" idea that positrons were electrons traveling backward in time helped lead to his own Nobel Prize–winning work. Feynman just needed to peel the speculative layers off the concept and get to the testable heart of truth.

Well after their days together at Princeton as professor and student, Wheeler's tutelage clearly had a profound influence on

Feynman's thinking. From learning a subject by teaching it to expressing concepts in terms of diagrams, Feynman borrowed much from Wheeler's systematic way of tackling new topics. Wheeler's courage to switch directions suddenly and embark on unexpected pursuits was echoed in Feynman's similarly unpredictable shifts, from quantum electrodynamics, to superfluids, to partons, and finally to computers. Finally, Feynman learned much from Wheeler's emphasis on a warm family life, which Feynman embraced in his later years.

Naturally, Wheeler gained much from having such a brilliant former student to tap for thoughts and ideas. He often sent papers to Feynman for his candid reaction. He gave some of his students projects related to Feynman's theories and methods—for example Charles Misner's exploration of sum over histories in gravitation. Finally, given his relatively staid personal life, he vicariously enjoyed some of Feynman's wild adventures when hearing about the details.

JOHN WYLER AND THE TRANSMOGRIFICATION OF DESTINY

By the mid-1970s, Wheeler's quirky style of disseminating his wild ideas, including odd terminology and evocative diagrams, had become familiar not just to his many former students but also to the growing theoretical community studying general relativity. *Gravitation*, which became a gospel of Wheeleresque philosophy and methods, spread the word even further. Published in 1973 and quickly translated into many languages, the thick volume was absolutely brilliant and completely idiosyncratic. Over the decades since its release it has remained a classic.

By then, practically anyone in the general relativity community could recognize a Wheeleresque turn of phrase or hand-drawn image almost immediately. It was almost as if he had invented his own language, accompanied inevitably by his own style of scientific art.

Just as Bohr's soft and enigmatic speech was well known to the Copenhagen quantum community, Wheeler's style and habits had become legendary as well. Bohr had often been parodied; now it was Wheeler's turn.

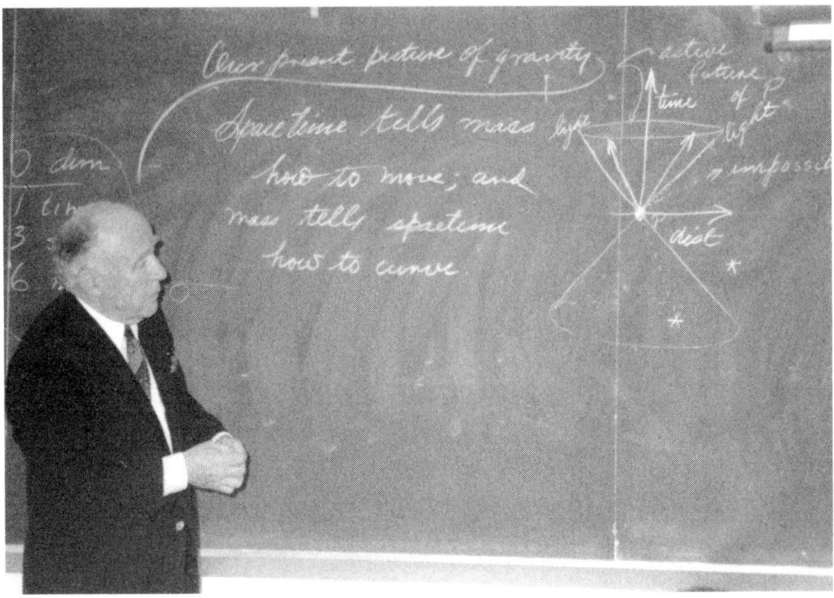

John Wheeler in 1986, lecturing about gravitation using a spacetime diagram (*Source:* Photograph by Karl Trappe, courtesy AIP Emilio Segrè Visual Archives, Wheeler Collection).

Members of the general relativity community were curious when they received copies in the mail of a typewritten manuscript by "John Archibald Wyler" of Princeton, titled "Rasputin, Science, and the Transmogrification of Destiny." As it went on, the paper got sillier and sillier, until it was completely ludicrous. It was written by physicist Bill Press, a student of Thorne at Caltech who had become a master at mimicking Wheeler's style, down to the lingo and unusual diagrams.

Here's a sample section of the clever parody, which was published in the journal *General Relativity and Gravitation* as an April Fools contribution in 1974: "Let the universe have a destiny. . . . Drop in a philospher [*sic*]. He makes a momentary disturbance on the void of destiny, then he too shrinks in cataclysmic annual repetition, and fades away. Drop in scholars of all disciplines and factions. They too are transmogrified until nothing more remains, only certitude. Call the resulting entity a 'black whole' and summarize the perfection of its final state. Say: 'A black whole has no peculiarities.'" The end of the article included a long list of acknowledgments, including shout-outs to "R. P. Funnyman," "F. Dicey," "S. W. Hacking," "C. W. Miser," "R. Pinarose," and so forth. Feynman often let papers pile up in his office, so even if he received a copy, he might not have seen the "Funnyman" reference.

Wheeler undoubtedly chuckled when he read the parody—sent to him by "Wyler." He had a marvelous sense of humor. He was well aware that others might find his terminology unconventional, but in a way that was the point. His colorful expressions, such as "sum over histories" and "wormhole," were far more memorable than the terminology they replaced. Although strictly speaking he didn't coin "black hole," his adoption of the expression brought it to public awareness. So he knew that being satirized was one price he would pay for being evocative. After all, that happened to Bohr too, in funny tributes such as the *Journal of Jocular Physics*.

THE COSMIC CODE

While Wheeler, Bekenstein, Hawking, and others were contemplating the role of information in black holes, Fredkin had already begun to envision an even grander role for meager bits: as elements of the computer code for the universe itself. Perhaps the continuity of time and space is an illusion. On the Planck scale, the cosmos could well be discrete, like the pixels on a television screen. Given the complexity generated by a simple computer program, such as the "Game of Life," perhaps the universe is governed on a quantum level by simple rules iterating zeros and ones as well. Maybe that basic algorithm generates all the laws of physics and other principles of nature.

Fredkin wasn't well known among physicists. He was brilliant but unconventional and never completed college. Nevertheless, Minsky took the chance of appointing him, on the basis of his talent and ideas, to MIT's computer science department. Funded in part by grants from the Department of Defense, both worked on artificial intelligence schemes.

Fredkin and Minsky met Feynman for the first time on a lark. While in Pasadena for a professional venture, they had some extra time and decided to call scientists in the area whom they admired. They got out the phone book. First they tried Nobel Prize–winning chemist Linus Pauling. Ring, ring, no answer. Next, on Minsky's suggestion, they dialed Feynman's number. Much to their delight, he picked up, and they had a terrific conversation. Even though he didn't know them, he invited them over to his house that evening, where they talked at length about computers and physics. They chatted late into the night. That marked the start of a fruitful connection.

In 1974, hoping to learn more about computation, Feynman invited Fredkin to Caltech as a visiting researcher in his department. It was a productive year. Fredkin picked up many insights from Feynman about quantum mechanics. He also solved a problem that had wracked his brain for years: designing time-reversible computer algorithms that worked the same forward and backward. If particle physics (except for certain neutral kaon decays) is time symmetric, so should be any digital processing schemes that modeled it. Fredkin's reversible system became known as the "Fredkin gate." As it turned out, Charles Bennett of IBM had already introduced a form of reversible computing, but Fredkin's device was still an achievement. Fredkin's visit in turn gave Feynman a window into the emerging field of artificial intelligence.

DO NOT GO GENTLE

In 1976, Wheeler reached retirement age at Princeton. At that point, he was far from ready to leave academic life. There were too many intriguing questions to explore. While assuming the title professor emeritus, he also accepted the generous offer of a postretirement professorship from the University of Texas, Austin (where the DeWitts had relocated). He was also appointed director of a new Center for Theoretical Physics. Despite having many friends in Princeton, he and Janette were eager to begin a new phase of their life together in the Lone Star State.

Feynman, in the meanwhile, was still having the time of his life in Southern California. He and Gweneth bought a Dodge Tradesman Maxivan and had its brown and yellow exterior custom-painted with Feynman diagrams. They wanted a custom license plate to go along with its motif. Restricted to six letters, they chose "Qantum." The van was perfect for camping trips and jaunts to a beach house they owned.

Feynman delighted in spending time with his two children. Michelle brought out his sweetness and fatherly devotion. He'd bring her various stuffed animals each night for bedtime, and she'd get to choose just the right one. Then he'd pretend to be a human radio, with her tweaking his nose to find each "station"—prompting him to make up songs in different styles. They'd laugh and laugh at their own clever silliness.

One time, when Michelle was somewhat older, a friend of hers whose family was going on vacation dropped off their pet boa constrictor at the Feynman house for a month of "babysitting." Freeman Dyson, on what turned out to be his last visit to the house, humorously recalled the ensuing pandemonium. Much to Feynman's consternation, the snake's diet was live mice. After he managed to procure some, he was dismayed to see that the boa constrictor was too languid to catch its prey. Instead, the mice began nibbling on the snake's skin, disfiguring it. Feynman had to stand guard to protect the pet. Michelle's friend's parents were very upset when they returned from vacation to find their "baby" with nibble marks. Never again, Feynman swore. More conventionally, Feynman had several dogs and loved to teach them tricks.

With Carl, then a teenager, Feynman enjoyed exchanging witticisms around the family dinner table, much as he had done with his own mother years earlier. She was well and living near them, often accompanying them to the beach. Those were happy, memorable times for all of them.

Around that time, on at least two different occasions, Wheeler visited Southern California and dropped by the Feynmans' house for dinner. As Carl later recalled, "I loved *Gravitation* as a child and was chuffed to meet another of the authors (I already knew Kip Thorne)."

A frequent dinner guest was Feynman's friend and drumming partner Ralph Leighton. They'd share a meal and retreat into his home studio to drum together jubilantly like wild men.

At one dinner in 1977, Feynman challenged his family and Leighton on their knowledge of geography. He asked if they thought they knew every country in the world. Leighton muttered, "Uh, sure." After pausing for dramatic effect, Feynman inquired, "What ever happened to Tannu Tuva?"

As a child, Feynman had seen stamps issued from that land and wondered where it was and what had happened to it. He and Leighton looked it up in the encyclopedia and realized that it had been incorporated into the Soviet Union, near the Chinese border. Researching its culture as much as they could, including its famed "throat singing," they thought the remote region sounded fascinating. Vowing to visit it together someday, they started contemplating ways to do so. As it turned out, Feynman would never make the trip.

In the summer of 1978, Feynman experienced the start of a decline in health that would limit his travel over the next ten years. Sadly, it would be his last decade. In early June, he received an invitation from Wheeler to participate in a planned symposium honoring the centenary of Albert Einstein's birth. Wheeler asked Feynman to contribute to a panel titled "Einstein and the Physics of the Future" or to speak instead about another topic of his choice. Feynman politely declined and jokingly reported to Wheeler that he had spoken with "Mr. X," who couldn't attend either.

Feynman was not in good shape at the time. He had begun to experience sharp pain in his abdomen. He went to a doctor, who diagnosed him with liposarcoma, a form of cancer. As the only reasonable option was surgery, he went under the knife. A football-sized tumor that had been pressing on his spleen and one of his kidneys was removed. He needed a number of months' rest at home until he was back on his feet.

On July 28, Wheeler sent Feynman a get-well letter. "Welcome home," he wrote, "from that experience . . . you rated only next to the misery of being President of CalTech. . . . Warm wishes for a speedy recovery."

In the envelope, Wheeler also included some reading material to aid in Feynman's recuperation. It was a piece titled "Law Without Law," which was part of his latest paper, "The Frontiers of Time." Prompted by teaching a class on quantum measurement theory, Wheeler had been giving renewed thought to the mysteries of interactions on the quantum scale.

WHO'S ON FIRST? WHAT'S ON SECOND? I DON'T KNOW, IT'S QUANTUM

That year, Wheeler concocted a brilliant thought exercise, called the "delayed-choice experiment," designed to elucidate the weirdness of quantum measurement. He imagined it as a way that he could have convinced Einstein that Bohr was right in their famous debates over whether quantum mechanics, with its probabilistic "dice rolling," could be a complete physical theory (Bohr said yes; Einstein said no). Strangely enough, Wheeler's hypothetical construct went even further than Bohr's complementarity, which stated that an observer's choice before or during a measurement might affect whether a photon, electron, or other subatomic object demonstrated wave- or particle-like properties. Wheeler's variation imagined how a decision made in the future might retrospectively affect which aspect is displayed.

Wheeler's configuration was simple but clever. He envisioned a baseball diamond with mirrors placed on home plate and each of the three bases. These mirrors would be of two varieties: first and third base would have regular mirrors that reflected all light, but home plate and second base would have special half-silvered (semi-transparent) mirrors that reflected half the light and transmitted the other half. The second-base mirror would have the additional property of lowering into or rising out of the ground at the flick of a switch. Initially, it would rest in the lowered position.

Suppose the half-silvered mirror on home plate was placed so that its back was facing second base, and a beam of light was shone diagonally on that mirror in the direction of first base. Because of the special mirror's properties, half of the light would bounce off perpendicularly and head toward third base, while the other half would pass right through and continue onward to first base. Within a tiny fraction of a second, the light aimed at third base would hit the mirror placed there and bounce off toward right field. Similarly, the light targeted at first base would reflect toward left field. Detectors placed in right and left fields would record the expected quantum answer: 50 percent of the original light would be detected on the right and 50 percent on the left. As in Feynman's sum over histories, both outcomes would occur simultaneously with equal likelihood. The light would be spread out over two distinct regions—like the rippled patterns in the well-known double-slit experiment, designed to demonstrate quantum fuzziness.

Now, imagine what would happen if the second-base switch was flicked so that its half-silvered mirror rose to the level of the other mirrors. Initially everything would happen the same, with beams heading toward first and third base in equal measures. However, after they bounced, each beam would hit the second-base mirror. If oriented in a certain direction, the second-base mirror would divert all the light from first and third base out to right field (transmitted in the former case and reflected in the latter) and none of the light out to left field (because of a cancelling out of the two beams in a process called "destructive interference"). The result would be a completely different outcome: 100 percent right, 0 percent left. The light would be focused in one place, like a particle.

Light takes time to travel; its movement is certainly not instantaneous. Therefore picture a situation in which the original beam was turned on and off so quickly that only a handful of photons (or even a single photon) made it out: a confined light packet rather than a continuous flow. Further suppose that an observer near second base was instructed to flick its mirror switch after the original packet was transmitted but before it reached second base. Arbitrarily, the measurer would decide whether the second-base mirror should be raised or lowered. Based on a delayed choice, rendered after the experiment had already started, either outcome could be selected: wave- or particle-like.

For example, suppose a light packet was sent out with the intention of producing a wavelike interference pattern with two distinct peaks in right and left fields. But then the experimenter changed her mind and flicked the switch to raise the half-silvered second-base mirror. The light would hit it and deflect completely to right field, creating a particle-like outcome. How could the light, once released, "know" that it had to transform itself? Or did flicking the switch somehow retroactively affect the properties of the transmitted light? If that were possible, then quantum measurement could operate backward, as well as forward, in time. Experiments conducted in 1984 and 2007 have since confirmed Wheeler's clever hypothesis.

After even more thought, Wheeler went a step further and extended his delayed-choice experiment to the universe itself. Instead of a baseball diamond, he imagined a superluminous, incredibly distant object, such as a quasar (energetic galaxy in formation), standing in for home plate and two appropriately positioned galaxies serving as first and third base. Through the process of gravitational lensing (bending of light due to the warping of space), they'd each steer light from the quasar toward Earth, which would act like second base (where the decision to raise or lower a half-silvered diverting mirror is made) in the baseball scheme.

On Earth, astronomers could decide to point a telescope to either of the lensing galaxies. Alternatively, they could use a half-silvered mirror to combine light from both. Aiming telescopes at either galaxy would yield a particle-like solution in which the quasar image appeared as a dot. In contrast, using the mirror option would produce a wavelike solution in which the quasar image was smeared out in a rippled pattern. Hence, the astronomical observers, billions of years after the quasar's light was emitted, could choose whether to render it as a particle or as a wave. When complementarity meets cosmology, the results are odd indeed.

A FRIEND IN NEED

Once Feynman was able to travel, Wheeler began to invite him again to research conferences, offering him the widest possible range of subjects. Anything his prodigal former student contributed would no doubt be fascinating. Jaded about some of the speculative topics, Feynman often turned him down—but occasionally he'd accept.

In 1981, Feynman decided to attend one such conference in the Austin area, held at the Lakeway World of Tennis Resort and Spa on Lake Travis. He and many of the other physicists found the venue's pervasive tennis theme kitschy. Freeman Dyson, who also attended, recalled that the swimming pool was shaped like a tennis racket. As it turned out, this would be Feynman and Dyson's last meeting in person.

When Feynman opened the door to his hotel room, he was flabbergasted. The majestic suite was way too fancy-schmancy for his spartan tastes—a complete waste of money, he thought. The management couldn't find a smaller, cheaper room for him; the hotel was sold out. So he decided to spurn the posh accommodation and sleep outside. The climate was desertlike and fairly warm, so he didn't think it would be a big deal. Once night fell, however, the temperature dropped, and he started to shiver. He grabbed a sweater out of his suitcase and covered up as much as he could.

Asked by a reporter for the local newspaper the *Austin American-Statesman* why a well-paid Caltech professor and Nobel laureate would sleep like a homeless person, Feynman replied, "I'm a big fool, but I enjoy life."

Feynman didn't stay out in the cold for long. Feeling sorry for his old friend, Wheeler invited him to lodge at his house. Feynman was most grateful. He told the reporter, "One of the biggest regrets of my life is that I'm not as nice as [Wheeler] is. I feel bad because I don't invite students to my house much or have the natural relations that he has with students."

THE BEAT OF THE BONGOS, THE ROAR OF THE CROWD

Despite his expressed feeling that Wheeler was nicer to students than he was, in truth Feynman and the Caltech undergraduates had a multidecade love affair. Feynman relished hanging out with and impressing young people, who, in turn, treated him as if he were Zeus on Mount Olympus. He was active in the university's orientation program, called "Frosh Camp," participating with freshmen in a variety of recreational activities and other challenges, enabling them to get to know him practically as soon as they arrived on campus.

When Shirley Marneus, director of Caltech's theater program, needed a bongo drummer for a production of *Guys and Dolls*, the student producer let her know about Feynman's talents. At that point, she hadn't even heard of him but thought inviting him to participate sounded like a good idea. By the time he accepted, she found out he was a Nobel Prize winner. Upon meeting him, in deference to his status, she called him "Dr. Feynman." He immediately insisted, "I'm Dick." He gladly took direction and had two roles in the play: bongo player and a voiceover during a backroom scene involving a game of craps. The audience loved it.

Feynman was hooked, finding the nuances of acting fascinating and enjoying being just one of the players. It gave him a chance to shed his image for a while and be someone else. He gladly took direction from Marneus, appreciating her expert advice on how to bring plays to life. The following year she gave him a bigger role: gangster Frankie Scarpini in a production of the musical *Fiorello!* Students laughed hard when he walked out on stage and hammed it up in goofy costumes.

Marneus had expected that a man of his stature would have wanted to spend very little time rehearsing. On the contrary, he would often stay for the full rehearsals, lasting several hours. When he wasn't called upon to perform, he'd frequently sit in the aisles and help students with their physics homework or work out problems with them on a blackboard backstage. Marneus found him extremely charming and helpful.

Feynman's former student and collaborator Al Hibbs held annual themed costume parties on April Fools' Day. These offered Feynman additional opportunities to dress up and act silly. He was outfitted on various occasions as a Ladakhi monk, God (sporting a long, gray beard), and Queen Elizabeth II. Marneus was amused when for an astronomy theme, he arrived in conventional clothes and called himself "Sirius."

She also saw Feynman's stubborn side when once, at a campus event, a visitor walked up to him with a copy of his book *QED*, a primer on quantum electrodynamics based on lectures he gave in 1979, and asked him for an autograph. Feynman was about to oblige, when the visitor said the book was so good it should be required reading for high school students. Recoiling from the suggestion, Feynman scolded him for trying to force a book on anyone and refused to sign it. The visitor pleaded, but Feynman kept saying no. Seeing the man practically in tears, Marneus finally persuaded Feynman to change his mind once more and autograph the book.

In October 1981, several years after his first operation, Feynman received bleak news. His cancer had returned and spread around his intestines. His only option was much more radical surgery to remove the cancerous tissue and much of the flesh around it. The operation, which lasted ten hours, proceeded relatively uneventfully and seemed a success, until a major artery burst near Feynman's heart as the surgeon was stitching him up. He lost so much blood that an urgent call went out for donations. Luckily, in a matter of hours, hundreds of volunteers, including many students, had answered the call, enabling the beloved physicist to recover. Feynman emerged a weaker man, having had a great deal of tissue removed, but ever so grateful that he made it through the procedure and could return to campus and family life.

Feynman thought, at first, that because of his fragile health he'd have to miss the next musical production: *South Pacific*. He was so depressed that his family insisted he speak with Marneus about a role. She suggested that he take on a minor part as the island chief of Bali Ha'i, surrounded by dancers and drummers. It would require wearing a colorful costume, complete with a large headdress, and issuing a few commands in Tahitian.

"But Shirl," Feynman responded. "I have a scar from the surgery." The scar was prominent on his belly, which the costume would expose. He certainly didn't want the audience to feel sorry for him.

Marneus gazed at him intently, envisioning him already as the chief. "You have that scar because you dove into the water to rescue the pearls. You fought off a shark, which bit you. You floated to the surface. The maidens lifted you into the canoe and covered you with flower petals. Then the village made you chief of Bali Ha'i."

"No kidding? Is that what happened? Then I guess I better do it!" said Feynman, imagining himself as a bold warrior rather than a recent patient. He never complained to her about the scar again.

When it was showtime, Feynman snoozed backstage until he was due on stage. He mustered his energy and put his soul into the performance. The audience was astonished and elated.

Marneus remembered the stunned reaction of the audience when, so soon after life-threatening surgery, Feynman first walked out on stage. "There was a moment of shocked silence. Then they screamed and were all on their feet applauding. He was really loved!"

Feynman would continue to act in numerous other plays, taking on roles as diverse as the Sewer King in *The Madwoman of Chaillot*, in which he doled out wry advice, to a custodian in a kickline in *How to Succeed in Business Without Really Trying*. He was simply a natural performer.

ARTIFICIAL MINDS

Feynman was delighted when MIT, his undergraduate alma mater, accepted Carl into its computer science program, and Carl decided to go there. Intrigued by artificial intelligence and the work of Fredkin and Minsky (who both worked there), Feynman thought it would be a great opportunity for his son. Throughout Carl's stay there, Feynman's interest in the fundamentals of computation continued to rise.

In May 1981, Feynman gave an influential speech, "Simulating Physics with Computers," in which he introduced the notion of quantum computing. He began the talk by acknowledging Fredkin's influence, to which he attributed his growing interest in the field. Then, starting with the concept of simple digital systems such as cellular automata, he explained how classical physics, in its deterministic form, might be simulated. The breakthroughs in reversible computation, he emphasized, were key to such simulations, because classical physics is time reversible.

For nondeterministic systems, probability could be built into the mechanisms, much as it is programmed into the workings of a slot machine at a casino. However, for realistic models of quantum systems, Feynman cautioned, standard automata and ordinary computers would not be enough. Reproducing the weirdness of quantum mechanics would require quantum computers, based on superpositions of states. He suggested using either electrons with superpositions of up and down spin states or photons with combinations of clockwise and counterclockwise polarizations as binary quantum elements. Such quantum generalizations of bits would become widely known as "quantum bits," or "qubits," a term often attributed to Benjamin Schumacher, who studied under Wheeler.

These quantum bits could be assembled into lattices, much like cellular automata, with each cell interacting with its nearest neighbors according to the rules of quantum dynamics. Such devices would bring sum over histories into the cybernetic realm, relying on nature's uncertainty to convey a broader spectrum of information, until a measurement was taken to resolve the superposition of quantum states into one of its components and yield the final result. Instead of taking one linear path to the answer, it would try out all possible options simultaneously, saving considerable time. It's like a maze with a lot of rats all seeking a piece of cheese; chances are they'd find it very quickly. Amazingly, four decades after Feynman proposed sum over histories, he was still finding new uses for the concept.

Feynman continued to take a strong interest in his son's studies at MIT. Through his undergraduate research at Minsky's Artificial Intelligence Lab, Carl had become involved in schemes for parallel processing: arrays of computer processors working in tandem for faster and more effective computation. In 1983, W. Daniel Hillis, a graduate student Carl had worked with, decided to found a company called Thinking Machines Corporation to design and manufacture a new generation of computers, called "connection machines," with a million parallel processors each.

Carl brought Hillis to meet his dad during a visit back home. While Feynman initially reacted to Hillis's plans with skepticism, he soon warmed to the idea. Hillis was astounded when Feynman—by then well enough to travel—volunteered to work for some time at the Boston area start-up. The business took off, and Carl became active in the company. Several years later Feynman would enthusiastically report, "A year ago I would have told you that the uses of a massively parallel computer are very limited. Now, it's getting harder and harder to find something that it can't do."

QUBITS AND SUPERSTRINGS

With black hole information theory still on his mind, Wheeler continued to preach "it from bit" to anyone who would listen. Although both he and Feynman focused on the place of binary computation in the universe, their differing approaches persisted. In general, Wheeler was the dreamer and Feynman the doer. Wheeler looked to the stars, the past, and the future, while Feynman considered how to make things work on Earth in the here and now.

In 1985, many theoretical physicists were excited about a new prospect for a unified quantum theory of gravitation and the other forces, called "superstring theory." Developed by Michael Green of the University of London and John Schwarz of Caltech, based on the ideas of many others, it had several unusual elements. First, it replaced point particles, such as quarks and electrons, with minuscule vibrating strands of energy, on the order of the Planck length. Because of their finite size, infinite terms in the field theory became finite as well, eliminating the need for renormalization. Also, it relied on a new symmetry between fermions, the components of matter, and bosons, the carriers of force, which could transform one into the other. Perhaps most surprisingly, it made sense mathematically in only ten or more dimensions. As observable spacetime only has four, the other six dimensions would be curled into a tight geometry on the order of the Planck length and hence unobservable.

A number of prominent theoreticians, frustrated with the lack of progress in quantizing gravity using standard methods (generalizations of quantum electrodynamics), turned to superstring theory as a promising avenue of exploration. However, Wheeler and Feynman were both dubious for different reasons. Wheeler deemed it not far-reaching enough; Feynman found it sparse on evidence.

"We should be looking at broader questions," Wheeler later said. "How come existence? How come quantum? I remember one colleague went to a lecture about string theory. He described it to me as like a Presbyterian minister preaching the Gospel."

"I had noticed when I was younger that a lot of old men in the field couldn't understand new ideas very well . . . such as Einstein not being able to take quantum mechanics," said Feynman. "I'm an old man now, and these are new ideas, and they look crazy to me, and they look like they're on the wrong track."

An article published that year by David Deutsch, "Quantum Theory, the Church-Turing Principle and the Universal Quantum Computer," presented ideas far closer to Wheeler and Feynman's interests. Deutsch showed how to generalize deterministic Turing machines into universal quantum computers based on qubits. He showed how quantum parallel processing could be faster than standard linear algorithms. Finally, he argued that Hugh Everett III's many worlds interpretation of quantum mechanics was the most logically consistent way of describing the operations of such a device.

Deutsch was not alone in championing Everett's interpretation. German physicist H. Dieter Zeh explored its implications in a variation called the "many minds interpretation," proposed in 1970. Zeh posited that the observer himself doesn't split upon observation but instead remains in a superposition of states along with what is being measured. Instead of collapsing, his wave function becomes entangled (linked into the same quantum state) with that of the observed quantum system. Why then does he perceive a definite measurement rather than a medley of possibilities? Because, according to Zeh, his mental state bifurcates into different alternatives, each with a separate, definitive conclusion. Since a body only has one mind guiding it, the other choices exist but are rendered inoperative.

Zeh helped develop another idea related to the many minds concept, called "decoherence." Wojczek Zurek, a student of Wheeler's at the University of Texas, was another of its prominent developers. Decoherence posits that upon each quantum measurement, a system becomes entangled with its environment. Because of that entanglement, within a brief period, its superposition decays into a particular definitive state, much as a tree, caused by the wind to sway more and more in one direction, eventually topples over. Only minute systems isolated from their environments might remain for long periods in superpositions. For large enough systems, exposure to the environment is constant and inevitable. Thus they remain in definitive states, rather than superpositions, and are called "classical."

THE SEEING I AND THE SELF-EXCITED CIRCUIT

Wheeler's own interactions with his inventive students swayed him more and more in the direction of "everything is information." He largely put his general relativistic interests aside in favor of quantum information theory. His application of the delayed-choice experiment to the universe itself steered him increasingly toward philosophical pursuits, much as had happened to his mentor Bohr. "Philosophy is too important to be left to the philosophers," he once said.

Wheeler called his new philosophy the "participatory anthropic principle." Like Bohr's complementarity, it emphasized the role of the observer. Curiously enough, however, because of delayed choice, that observer has the power to shape the past as well as the future. An example is the astronomer adding a half-silver mirror to his telescope and rendering photons from an ancient quasar wave- instead of particle-like. From his prior work in geometrodynamics and quantum foam, Wheeler believed that affecting the wave functions of structures in the past could shape the destiny of the universe itself. Perhaps human observation, therefore, molded the primordial universe in such a way that it evolved the capacity to sustain life. Consequently our species today, with its far-reaching observational powers extending back into the distant past, in some sense would have created the conditions for its own existence. Drawing upon an electronics analogy, Wheeler dubbed this idea the "self-excited circuit" and sketched a memorable drawing of it as a U-shaped object possessing an eye on one side gazing at its own past on the other.

"The universe does not exist 'out there,' independent of us," Wheeler once wrote. "We are inescapably involved in bringing about that which appears to be happening. We are not only observers. We are participators. In some strange sense, this is a participatory universe."

Wheeler often discussed "Twenty Questions: The Surprise Version," a game that elucidated the concept of observations creating something new. In it, a group of friends secretly agrees that they are going to play the classic game of Twenty Questions with a twist: at the beginning of play, none would have a particular word in mind. The questioner (who wasn't in the room to hear about their conspiracy) wouldn't be told of their ruse. Instead, after the questioner posed the questions, each would listen to the others' answers and make sure that theirs was consistent with all the previous ones. Naturally that would narrow down the possibilities more and more.

For example, suppose the questioner asks, "Is it a physicist?"

Without having anyone in mind, the first player says, "Yes."

Thinking it might be Einstein, she asks, "Is it someone who plays the violin?"

The second player says, "No."

Next, she asks, "Is it someone who plays any instrument at all?"

The third player says, "Yes."

She then asks, "Is it someone born in Europe?"

The fourth player says, "No."

On a whim she asks, "Is it someone who plays the drums?"

The fifth player says, "Yes."

That narrows the possibilities down quite a lot. The interrogation goes on, and finally, when she is about to run out of questions, she asks, "Is it Feynman?"

Even though the group didn't have Feynman in mind in the beginning, the last player couldn't think of anyone (or anything) else matching the description delineated by all the other answers. So after a long pause, he had to answer "Yes," generating the answer "Feynman" from the results of all the "observational outcomes."

A deep connection between Wheeler's game and "it from bit" was that all the questions had binary answers: yes or no, equivalent to one or zero. Hence not only was the answer created by questions but also it could be represented by a binary string of all the answers to the questions. Similarly, the binary switch settings in the delayed-choice experiment—arbitrarily raising or lowering the key mirror that determined wave- or particle-like behavior—encoded its outcome. Applying the delayed-choice experiment to the universe itself, as did Wheeler, one might imagine encoding its properties retroactively by means of a series of binary decisions about what types of cosmological measurements to take.

When asked in an interview about Wheeler's notions regarding how the laws of the universe came into being, Feynman declined to weigh in, conceding the topic was too speculative for his taste. In the same interview he also refused to opine about whether or not the many worlds interpretation could be correct. His focus remained on more pragmatic concerns.

"All I'm interested in," Feynman said, "is trying to find a set of rules which would agree with the behavior of nature, and not try to go very far beyond that. I find most philosophical discussions are psychologically useful but, in the end, when you look back historically on what was being said, and being said with such vigor, it's almost always—to a degree—nonsense!"

RETURN TO PRINCETON

Wheeler's mind remained incredibly active, but age had started to take its toll. In April 1986, he had open-heart, triple-bypass surgery at Seton Medical Center. The harrowing experience got him thinking about his own mortality. At that time, the procedure was very risky, requiring that his heart be stopped and packed in ice for two hours. Following a successful procedure, he needed to stay off his feet for about two months. Luckily, Janette was extremely supportive.

By June, he was feeling much better. He felt like he had a new lease on life. In a moment of reminiscence, he wrote to Feynman recalling their adventures together: "Physics was exciting then. I find it even more exciting now, and one of these days I'm going to barge in again on you and chew the fat about the physics of information."

Around that time, Feynman's name was much in the news when, in response to the *Challenger* space shuttle disaster earlier that year, he was invited to serve on the Rogers Commission examining the causes of the tragedy. Characteristically, wanting to arrive at his own unbiased conclusions, he conducted an independent investigation. His focus turned to the rubber "O-rings" used to seal joints in the shuttle's rocket boosters. After examining their properties, he concluded that they weren't resilient enough to temperature changes. At the hearings, he dipped an O-ring into ice water and demonstrated its lack of resilience. After the commission's report was prepared, he found it too noncommittal and issued a much more damning critique of his own, included as an appendix. Detailing the mistakes made, including officials' failure to anticipate the formation of cracks in various systems, it ended with an admonition: "For a successful technology, reality must take precedence over public relations, for nature cannot be fooled."

Wheeler's mind kept whirling with offbeat notions. In August, he sent Feynman a speculative article that he had recently completed, "How Come the Quantum?" He attached a note advising Feynman of its outrageousness: "Didn't I inherit from you the faculty of coming up with crazy ideas?"

Wheeler's speculative notions, always a bit on the fringes, had become almost unfathomably abstract. They were so philosophical that no one could imagine how to test them. What kind of experimental data could resolve questions such as "How come existence?"

Yet Wheeler had no desire to be known as a New Age guru or pseudoscientist. He complained vehemently, for example, when at a meeting of the American Assocation for the Advancement of Science, he found himself seated on a panel with parapsychologists. Neither did Feynman want such associations, but in 1984 he did give a talk titled "Tiny Machines," updating his views on nanotechnology, at the Esalen Institute in Big Sur, California, a New Age haven for hot-tubbers. He also experimented with flotation tanks, seeing what isolation and sensory deprivation would do to his thoughts. Wheeler would trade a hot tub or flotation tank any day for a quiet, rocky, secluded beach in Maine.

A portrayal in the September 1986 issue of *Reader's Digest*, "Inside the Mind of John Wheeler," exaggerated some of his views and brought him much unwanted attention from mystical believers. The piece implied that he had found a link between science and religion. Consequently, letters began to pour in from dozens of would-be devotees from all over the world—as if he were the maharishi of physics. He let the fan mail—undoubtedly full of crackpot theories vying for his notice—pile up and decided to ignore the whole incident.

Despite feeling energized, Wheeler felt, at the age of seventy-five, that it was time to retire his University of Texas position. Ten years in Austin had been an incredibly productive experience, steering him to hitherto uncharted territory. Yet, like Odysseus after his expeditions, he was ready to go home—which for the Wheelers meant the East Coast and particularly the Princeton region. After finding a retirement community within a short drive of Princeton University and arranging for an office at Jadwin Hall (the new home for physics at the university), to which he was entitled as a professor emeritus, they were all set. They moved at the end of February 1987.

WORMHOLES AS PORTALS TO THE PAST

In the final decades of his career, Wheeler mentioned black holes far more often than wormholes in his public talks and writings. Guiding him was a focus on the viable, which, despite his far-reaching speculations, he considered a critical litmus test. While by then astronomers had identified many black hole candidates, wormholes had remained hypothetical constructs without realistic, stable solutions. Physics journals hardly mentioned them.

Nevertheless, when in the mid-1980s astronomer and science writer Carl Sagan asked Kip Thorne if he knew of any credible schemes for interstellar travel (for use in Sagan's novel *Contact*), Thorne decided to dust off wormholes and see if he and his student Michael Morris could employ them (theoretically speaking) as spatial shortcuts. Rather than minuscule wormholes embedded within the spacetime foam, as Wheeler had pictured them, Morris and Thorne searched for hypothetical objects large enough and stable enough for spaceships to traverse safely to reach otherwise remote parts of the cosmos.

Soon Morris and Thorne identified the key ingredient that could potentially make traversable wormholes possible: matter conjectured to have negative mass. By sculpting such material, along with ordinary positive mass substances, into particular arrangements, they could craft wormholes with entranceways spacious and sturdy enough to allow safe, speedy passage. In principle, an astronaut could enter one "mouth" (entranceway) of the wormhole, pass through its "throat" (connecting region), emerge unscathed through a second mouth a reasonably short time later, and explore a far-flung region of the universe.

The research team fully recognized that their wormhole transportation scheme would require technologies far beyond present capabilities. For one thing, its overall mass would be comparable to that of a galaxy. Only an immensely powerful, advanced civilization would have the capacity to assemble such a massive structure. Moreover, there is no known material with negative mass, one of the critical wormhole ingredients. Despite such caveats, the article Morris and Thorne published about their scheme generated considerable interest in the gravitational physics community and inspired numerous other papers on the subject.

Soon after completing his first paper with Morris, Thorne invited another of his students, Ulvi Yurtsever, to join their collaboration and develop a second project related to wormhole time travel. The group demonstrated how it might be possible to manipulate wormholes in such a way that they would allow journeys to the past. It made Kurt Gödel's dream of loops through space, inducing backward voyages in time (which in his version required a rotating universe), much more credible.

In relativity, traveling to the future is relatively straightforward. Assuming the technological challenges could be ironed out, simply hop on a spaceship and travel close to light speed. Due to time dilation, your internal clock would run much slower than that of those you left behind. Therefore, if you returned to Earth, your friends and family members would have aged faster than you did. In other words, you'd have leaped ahead of them in time. The faster your journey, the greater your time dilation, and the farther into the future you would have traversed.

Traveling to the past is much trickier. You'd have to circumvent the law of cause and effect and reverse the hands of time. Yet, as Thorne's team demonstrated, if an advanced civilization could create a traversable wormhole and speed up one of its mouths close to the velocity of light, a spaceship could journey into that mouth, through the throat, and out the other mouth to voyage backward in time. The scheme relies on time dilation of the first mouth compared to the second, rendering the former farther in the future than the latter.

For example, suppose an alien civilization, back in our year 1938, constructed a wormhole and sped up one of its mouths such that in a single year of the wormhole's time, it aged one hundred years of our time. That mouth would "live" in 2038. The other mouth, at rest compared to the first one, having aged one year both in its time and ours, would "live" in 1939. Imagine, for the sake of argument, that both mouths were close enough to Earth that a terrestrial astronaut could access each of them and return to our planet in a reasonable time. Consequently, an intrepid voyager could enter the first mouth in the year 2038, pass through the wormhole's throat, and venture back to 1939 via the second mouth. If he then returned to Earth, he could visit the very moment that Feynman met Wheeler!

If you've read a lot of science fiction, you're probably thinking at this point about time travel paradoxes. For instance, what if you ventured back to Princeton in 1939 and made sure that Feynman was assigned as Eugene Wigner's teaching assistant instead of Wheeler's? Maybe particle physics would have taken a different course. To avoid disrupting history, Thorne and his group proposed that backward journeys in time must be self-consistent. In other words, anything that happened in the past must coincide with the known course of events. That is, while you could play a role in history, such as offering advice to the young Feynman, you couldn't actually change it. Your escapades in the past would simply be a chunk of the concrete edifice of time.

Wormholes remain hypothetical creations. Observational astronomers are far more interested in known celestial bodies and the events that mark their development. Of such occurrences, one of the most exciting is the burst of a supernova.

FEYNMAN'S SUPERNOVA

A massive star marks the end of its life with a fantastic fireworks display. For a brief interval, supernova bursts release extraordinary amounts of energy, more so than that of an entire galaxy. Such blasts include photons of various frequencies, neutrinos, gravitational waves, and material ejected from the star. The energetic photons heat up the interstellar gas around the exploded star, combining with the ejected material to create colorful patterns known as "supernova remnants."

Supernova blasts in any given galaxy, including the Milky Way, are rare occurrences. Catastrophic events close enough to us to be visible to the naked eye happen only once every few centuries. For that reason, astronomers were excited when, on February 23, 1987, they detected a supernova explosion in the Large Magellanic Cloud, a satellite galaxy of our own, about a million trillion miles away. A few hours after its gargantuan release of light energy reached our planet, a barrage of invisible neutrinos arrived. Detection of those, using special facilities deep underground, marked the dawn of the era of neutrino astronomy.

Joe Weber, collecting data using his apparatus at the University of Maryland, and physicist Edoardo Amaldi, employing a similar instrument in Rome, asserted that they had found joint evidence of gravitational waves from the explosion. Both researchers claimed to have detected similar signals around the same time. However, most astronomers maintained that their instruments weren't sensitive enough by a long shot. Therefore, their results weren't widely accepted. If the Laser Interferometer Gravitational-Wave Observatory had been online at the time, it would have offered the perfect opportunity to search for such signals.

Shortly after the supernova discovery, while delivering a guest lecture at a physics course for Caltech freshmen, Feynman said, "Tycho Brahe had his supernova, and Kepler had his. Then there weren't any for 400 years. Now I have mine." He likely meant that he felt lucky to have lived during the time of such a rare event—especially because he had come so close to death earlier.

Feynman's career had been far more than a sudden supernova burst; it glowed consistently and strongly for many decades. However, the stresses of his recurrent, spreading cancer were starting to wear him down physically. In October 1986, he'd gone under the knife once more and still wasn't back in shape. He seemed to have aged considerably in only a few years. Nonetheless, he remained uncomplaining and optimistic, focusing much of his attention on planning with Leighton their long-discussed trip to Tuva.

His final and most taxing operation took place in October 1987. With the excision of a significant amount of malignant tissue, as well as some of the healthy tissue around it, his kidneys failed, and he needed dialysis. Afterward, keeping up his daily activities became a major challenge, as he was very weak and in considerable pain. Nevertheless, he felt obliged to teach the graduate course in elementary particle theory to which he had been assigned.

Despite the surgery, the cancer returned a few months later—this time in inoperable form. On February 3, 1988, Feynman, feeling very ill, was admitted to the UCLA Medical Center. Given his dire condition, he asked for no extraordinary measures, just a comfortable finale to his full life. He felt that he had already contributed the best of his ideas to the world, so a longer life wasn't essential. He had just completed a second volume of his humorous recollections—the first had been a major bestseller—which focused his mind on the past.

At one point Jerry Zorthian and his wife paid a visit. Normally, Feynman would have clowned around, but he was in too much pain. Memories of Arline's final years and untimely death flooded back, and he began to cry. Seeing how exhausted his friend was, Jerry said good-bye for the last time. Feynman put on a brave face, urging him to go out and have fun.

A few days later, on February 15, Feynman spent his final hours with Gweneth, his sister Joan, and his cousin Frances Lewine by his side. As it turned out, the vice president of the Soviet Academy of Sciences had just mailed him an invitation to visit Russia and Tuva, but it would arrive too late.

Since the Zorthians' visit, he had been in and out of a coma. During a brief interval of consciousness, with barely the strength to speak, he slowly voiced his last words: "I'd hate to die twice. It's so boring."

That day, Caltech students hung a large banner from the tall Millikan Library building: "We love you, Dick." He was their legend, their hero, their scientific Houdini. Perhaps they hoped that somehow he could magically undo the ravages of time and reappear back on campus as a healthy man—like he had miraculously showed up in the campus musicals. If anyone could crack the secret of surviving against all odds . . .

HOW COME EXISTENCE?

How come existence? How come death? Why did Feynman's time on Earth end so soon?

In his final decades, Wheeler faced the sobering realization that he had outlived some of his prominent students. Hugh Everett died in 1982 of a heart attack at the age of fifty-one. At least he had lived long enough to see his work celebrated, thanks to DeWitt. Since then, Everett's many worlds interpretation has reached even wider audiences through media references, such as the BBC documentary *Parallel Lives, Parallel Worlds*.

Peter Putnam, with whom Wheeler felt a deep emotional bond, had left the field of physics soon after graduating. After some time teaching philosophy at Union Theological Seminary, he moved to Houma, Louisiana, to provide legal services for impoverished families, while working night shifts as a janitor. In the end, he was dirt poor himself. While cycling one evening in 1987, he was hit by a drunk driver and killed. Decades earlier, Peter's mother, Mildred Putnam, had donated a wonderful sculpture collection to Princeton, primarily to honor her other son (who died in battle) but also as a tribute to Wheeler's kindness. The John B. Putnam Jr. Memorial Collection still enlivens the campus.

Wheeler often reflected on those who were gone. He had so many fond memories of Feynman, he hardly knew where to begin. Nevertheless, he remained upbeat, valuing any time he had left on Earth to enjoy with Janette, along with their children and grandchildren. He remained active in the physics community, attending conferences, contributing papers, and meeting with young researchers. He divided his time between Princeton, where, aided by his assistant, Emily Bennett, he maintained an office as professor emeritus, and High Island, Maine, his summer retreat where other family members congregated. Quietly and generously, he contributed to efforts in understanding the history of physics. Many of his former students stayed in close touch with him, especially Ken Ford, with whom he wrote his autobiography.

Once he reached his nineties, he found that he needed to slow down and concentrate his efforts to a far greater extent. "How come existence?" was the question on which he chose to focus. There was, of course, no easy answer. Nevertheless, even when his health began to decline (he had a heart attack in 2001), he would go once or twice a week to his office to read his correspondence, to update himself on any physics news, and mainly to think.

On the morning of April 13, 2008, a quiet spring Sunday, John Archibald Wheeler died peacefully at his home at the age of ninety-six after a bout of pneumonia. His *New York Times* obituary quoted words by Freeman Dyson: "The poetic Wheeler is a prophet standing like Moses on the top of Mount Pisgah, looking out over the promised land that his people will one day inherit."

CONCLUSION: THE WAY OF THE LABYRINTH

The word "time" was not handed down from heaven as a gift from on high. The idea of time is a word invented by man—and if it has puzzlements connected with it, whose fault is it? It's our fault.

—John A. Wheeler, quoted in the film *A Brief History of Time* (1991)

You are in a maze of twisty little passages, all different.

—*Colossal Cave Adventure* (early interactive computer game by Will Crowther and Don Woods)

How come a conclusion? Why are we bringing this book to a close with a final chapter? What does that say about the nature of time?

In many ancient cultures, an ending was always just a beginning. Sagas were repeated over and over, word for word, to generation after generation of listeners. Life's demise always led to a new incarnation.

The rhythms of daily life and the motions of familiar astronomical bodies, such as the sun, moon, and planets, suggest that time runs in cycles. Their patterns repeat themselves again and again, with predictable results. Knowing exactly what will happen is reassuring. No wonder many people relish rituals, from religious ceremonies to annual holidays.

While cyclical time is comforting, linear time presents a rewarding challenge. Reaching the end of a book signifies a milestone: a creative effort completed. A linear tale, with an introduction, body, and conclusion, offers a sense of order and purpose. One might await the close with rapt anticipation, for better or worse.

Time seems to have many arrows. There is the cosmological arrow of the expanding universe, the thermodynamic arrow of nondecreasing entropy, the evolutionary arrow of growing complexity, the decay profile arrow of certain weak interaction processes, the psychological arrow of conscious awareness, and so forth. How these arrows are all connected remains a mystery.

Cyclical versus linear time has been the traditional contrast. Philosophical considerations steer many thinkers into choosing one or the other—for example, in debates about whether there was a time before the Big Bang or even a Big Bang at all. Ultimately scientists prefer to base their models on evidence rather than pure speculation. Clearly nature (at least on the familiar, classical level) has bits of both: some processes are repetitive, while others follow singular paths.

Today we encounter in our daily lives a third way of looking at time: as a labyrinth of myriad possibilities. In the modern information age, thanks to the Web and hypertext, we find ourselves in a maze of virtually unlimited complexity. The works of Jorge Luis Borges, Philip K. Dick, and many other speculative writers imagine time as a kaleidoscope of alternative realities interacting with each other. Now we find, due to the proliferation of online choices we make each day, that just as in such literary treatments, we are forever lost in a garden of forking paths.

The Internet, with its labyrinthine structure, is an emblem of science's move toward parallelism and away from cyclicality and linearity. The essential idea is that nothing is preordained to move in a circular, straight-line, or curved trajectory. Rather, the natural state is for every component of a system to interact in every possible way—a jumble of choices.

Only conservation laws and other physical bounds—for instance, conservation of electrical charge—restrict such interactions. Sometimes those get revised due to experimental evidence, which requires a rethinking of how to impose new restrictions.

Finally, Ariadne's thread through the labyrinth of choices emerges: an organizing principle. Such a selection mechanism reveals the optimal route through the realm of possibilities. Sometimes, as in classical physics, that is the one definitely chosen; other times, as in quantum physics, it sets the peak of the probability distribution. In the case of reading, a book—with an introduction, body, and, yes, a conclusion—might serve as an organizing agent for the subject it covers. The choices made by its author(s), editor(s), and so forth create a linear narrative that serves as a guide through the greater labyrinth of information.

As Richard Feynman realized early on, the prototype for all this is optics. Simplistically, we imagine that light travels in rays, bounces off mirrors, and bends through lenses because it is always tightly focused in a thin beam. But unless the light is an extremely narrow laser beam, that is not the real picture. Feynman's reading of Fermat's principle of least time informed him that in general such behavior constitutes just the crest of an unseen jumble of interfering wave patterns. Least time, as an organizing principle, brings order to the hodgepodge of light waves in space, resulting in light rays.

Feynman brilliantly applied the same concept of a labyrinth of interacting components—subject to conservation laws and an organizing principle to guide them—to the domain of elementary particles. As he would describe his overall methodology during a workshop at the Esalen Institute, "The game I play is a very interesting one. It's imagination, in a tight straightjacket, which is this: that it has to agree with the known laws of physics."

John Wheeler marveled at how beautifully Feynman's sum-over-histories approach encapsulated the distillation of the gamut of quantum possibilities down to a definitive result—linking quantum and classical in an unprecedented fashion. While particles and fields interact in every mode that is physically allowable, a weighted tally of their series of contacts yields what we actually observe. Wheeler's advocacy helped inspire other great physicists, such as Bryce DeWitt and Charles Misner, to explore possible models for quantum gravity. The questions Wheeler raised about quantum measurement steered Hugh Everett into proposing, as an alternative, his many worlds interpretation in which observers split along with the quantum systems they observe.

Feynman's diagrams depicting the possibilities in a sum over histories have become an essential lexicon for contemporary theorists. Extended to the weak and strong interactions, as well as electromagnetism, they have proved instrumental in the development of a Standard Model of particle physics. With its comprehensive description of the interplay between the forces (except for gravity) and known material constituents of nature, the Standard Model is one of the most successful physical explanations of all time.

Throughout his life, Wheeler aspired to understand the most fundamental components of the cosmos. He changed his mind on that issue several times in his career, starting with particles, venturing into fields and geometry, and finally delving into information. He also wanted to comprehend the organizing principles steering those components into recognizable patterns. Sum over histories, based on the principle of least action applied to quantum physics, was one such idea, but he also considered others. In the end, he became convinced that the answer had to do with a "self-excited circuit": a symbiosis between conscious observers and what they were observing, namely, the cosmic past. Somehow, through our looks back in time, we organized our own universe, from among the frothy possibilities of the quantum foam. Therefore, in Wheeler's mind, the questions of "How come existence?" and "How come the quantum?" became inextricably linked.

Today, while we celebrate the Standard Model, we recognize its limitations and wish to move beyond it. One of its glaring omissions: it doesn't include dark matter and dark energy, invisible components of the universe recognized during Wheeler's final decades but still unidentified. Dark matter is the hidden "glue" that keeps galaxies intact and binds them into clusters. Vera Rubin, who took courses with Feynman and Hans Bethe at Cornell in the late 1940s, demonstrated the need for such missing material in the 1960s and 1970s through galactic rotation studies conducted at the Carnegie Institution of Washington along with Kent Ford. In examining dozens of spiral galaxies, Rubin and Ford discovered that their outer stars revolved around their central hubs at a much faster rate than expected from the pull of visible matter alone. Consequently, much galactic material cannot be seen. Further astronomical observations have confirmed the presence of dark matter throughout the universe but not its identity.

Dark energy, the unknown propellant of accelerating cosmic expansion, is another great scientific mystery. As two teams of researchers discovered in the late 1990s, not only has space been expanding since the Big Bang but also its rate of growth has been speeding up. In 2011, team leaders Saul Perlmutter, Brian Schmidt, and Adam Riess received the Nobel Prize in Physics for their discovery. No one knows what causes space to enlarge at an ever-faster rate. Scientists are uncertain if the pace will pick up even further, slow down, or remain steady. Curiously, the cosmological constant, discarded by Albert Einstein after Edwin Hubble's 1929 discovery that galaxies are moving away from each other, has turned out to model well the effects of dark energy.

Searches for possible components of dark matter and dark energy are currently under way. If researchers identify such ingredients, they will likely need to modify the Standard Model to accommodate them. Possibilities for dark matter include axions, hypothetical particles proposed by physicist Frank Wilczek to explain why the strong interaction, unlike the weak, has *CP* invariance, as well as supersymmetric companions to ordinary particles. Certain hypothetical extensions of the Standard Model that endeavor to represent low energy limits of superstring theory predict the latter. The nature of dark energy is even trickier to discern, with few credible leads.

Another contemporary conundrum, leftover from the days of Feynman, Wheeler, and DeWitt, is why gravitation is such an oddball. Why is it so much weaker than the other interactions? How might it be described in a mathematically consistent fashion using the methods of quantum field theory?

The leading proposal today for a unified description of the natural forces, including gravitation, is M-theory, a generalization of superstring theory that includes vibrating energetic membranes, along with various configurations of strings (supersymmetric as well as not supersymmetric). Instead of point particles, its fundamental components are Planck-length-sized strings and membranes, interacting with each other in a variety of modes. They are mathematically consistent only in spaces of ten or eleven dimensions, with at least six of them curled up tightly into pretzel-like shapes called "Calabi-Yau manifolds." Theorists have modified Feynman diagrams to take into account extended objects and their possible interactions in such higher-dimensional spaces.

The biggest problem with M-theory is that it offers a mind-blowing range of possibilities for the properties of its components and the ways the Calabi-Yau manifold might be configured. For the latter, some theorists estimate around 10^{500} (one followed by five hundred zeroes) possibilities: an insanely complex labyrinth beyond even the wildest nightmares of Borges. Narrowing down the "landscape" of M-theory to include reality and nothing more would be a daunting process, requiring an extremely powerful selection rule. While Stanford physicist Leonard Susskind—who was a friend of Feynman—has proposed using the anthropic principle to do the honing, others have doubted that it has sufficient strength as a selection principle to rule out so many alternatives.

A related speculative idea is the concept of a "multiverse": an ensemble of more than one universe. Unlike Everett's many worlds, multiverses would exist in physical space, albeit in regions we cannot access. The concept took flight in the 1980s when physicist Andrei Linde proposed the concept of "chaotic inflation." In his scheme, a variation of Alan Guth's earlier inflationary universe model, the universe begins as a breeding ground for random quantum fluctuations in what is called a "scalar field." Particularly favorable fluctuations generate the seeds of bubble universes, which undergo ultrarapid intervals of expansion, called "inflationary eras." Stretching out space extremely quickly helps smooth out temperature differences, in line with the large-scale uniformity of the cosmic microwave backward, resolving the issue Misner identified when he developed the Mixmaster universe.

The idea of a multiverse leads to many weird possibilities. For example, another bubble universe could randomly produce a planet identical to Earth, but with minor differences, such as, for example, John F. Kennedy not being assassinated in 1963. In general, the concept lends itself well to "what if" alternative history scenarios.

Note that one doesn't need multiple bubble universes to posit duplicate or near-duplicate Earths. An infinite single universe will do just fine. The more planets in the universe, the greater the chances of Earth's course of development happening elsewhere. Maybe other versions of you, on planets very similar to our own, are just finishing up other versions of this book right now. Congratulations to all of you!

Now, we've reached the conclusion of the conclusion—the end of our winding journey through space and time. Our hunt for the ghosts of the past took many twists and turns, including close encounters with alter egos "Geon Wheeler," "John Wyler," "R. P. Funnyman," and the notorious "Mr. X." We've met bongo players, wild artists, solipsistic electrons, and a spouse who hated calculus. We've rested outside fancy hotels and inside seedy ones. The number of "crazy ideas" we've seen boggles the mind. All the while we've kept our sanity through a comforting, guiding principle: as sum over histories tells us, no matter how strange our path through spacetime, there exist many others that are even more bizarre.

EPILOGUE: ENCOUNTERS WITH WHEELER

The two protagonists of this book live on in the memories of those who worked with them, lived with them, studied under them, collaborated with them, and otherwise encountered them. Hundreds of students took Richard Feynman's "Physics X" course at Caltech over the years and still remember his madcap, colorful demonstrations and personal warmth. Many participated with him in the musicals or at least witnessed his drumming and silly costumes. Then there are the millions who watched his television presentations, on the BBC and elsewhere.

While I never personally met Feynman, I saw John Wheeler speak several times. I recall a talk he gave at an American Physical Society meeting during which, as he spoke about the death of his brother Joe, his voice cracked, and tears welled up in his eyes. It had remained one of the most painful memories of his life.

On another occasion, I was invited to an academic celebration of Wheeler's ninetieth birthday, titled "Science and Ultimate Reality," sponsored by the John Templeton Foundation and held near Princeton in early 2002. The list of speakers was phenomenal, covering a broad spectrum of his interests over his long and productive career. One of the highlights was Bryce DeWitt speaking at length about the many worlds interpretation. I later saw him chatting in French with his beloved Cécile. I believe that was one of the last conferences they attended together, before Bryce DeWitt's death two years later. A number of prominent young physicists—including Lisa Randall, Juan Maldacena, Lee Smolin, Max Tegmark, and many others—gave excellent talks about their research. For each talk and discussion, Wheeler sat in the front of the large lecture hall, taking it all in.

Later that year, I received a Guggenheim Fellowship to study the history of higher-dimensional theories in physics. When I decided to interview venerable physicists who might have been familiar with those developments, Wheeler's name leapt to mind. I wrote him a letter and was put in touch with his friend, collaborator, and former student Ken Ford. Ford helped me arrange a morning meeting with Wheeler, who still maintained an office in Princeton's Jadwin Hall. I was thrilled to receive several hours of his time.

While I was warned in advance that Wheeler's memory wasn't perfect, his recollections during our meeting were riveting. He spoke about what it was like to have Albert Einstein as a neighbor and to bring his students over to meet the renowned physicist. He mentioned how he had tried to convince him that Feynman's sum over histories made quantum theory more palatable. Einstein refused to budge in his opposition.

On a more humorous note, Wheeler said that his children's cat once wandered off to Einstein's house. Einstein called him to let him know. After he retrieved the cat, he asked it if it had learned anything about general relativity.

Wheeler's dry sense of humor was truly delightful. When I praised his book *Gravitation* (coauthored by Charles Misner and Kip Thorne), he handed me a copy in Chinese. "Well here's your chance to go on. You've read it in English, now you can read it in Chinese," Wheeler told me with a sly smile.

I could see why Wheeler's students and colleagues loved and respected him so much. He was a man of charm and grace, exceptionally polite and remarkably quick-witted. He thought deeply about how to live a life that was both meaningful to himself and of service to others. Meeting him was a magically transformative experience, as he was so warm-hearted and inspiring. By that point, his main research focus encompassed the meaning of life, or as he put it, "How come existence?"

Wheeler had just won the inaugural Einstein Prize for gravitational physics, along with Peter Bergmann. He had called Bergmann to congratulate him and left a message, but before they could speak on the phone, Bergmann passed away.

I wasn't sure if I'd ever see Wheeler again, but luckily another opportunity arose. In June 2004, Philadelphia held "The Big Nothing," an artistic celebration of emptiness and existence. It was just up Wheeler's alley. A clever group of artists at Temple University's Tyler Gallery decided to center their show on Wheeler's works, as documented in the American Philosophical Society archive of his donated material. They called the show the "Mixmaster Universe."

Thrilled to learn about the exhibition, I volunteered to contribute a biographical essay about Wheeler to be distributed to attendees. I also pointed out that Misner had coined the term "Mixmaster universe" and suggested that they invite him too. He saw the show and seemed intrigued by the variety of artistic interpretations of his and Wheeler's work.

Much more recently, I had the chance to meet Drs. Jamie and Jenette Wheeler (John's son and daughter-in-law), first at a talk by science writer Amanda Gefter and shortly thereafter at a discussion of Feynman that I led at the Lantern Theater after its production of the Feynman play *QED*. In the latter encounter, Jamie Wheeler gave a lively account of Feynman's soup can demonstration, which he recalled from childhood. History is always so much more vivid when conveyed via firsthand accounts.

I shall always cherish my opportunity to chat with John Wheeler and my other encounters with him and his colleagues. His spirit persists in the lives of everyone he has touched with his insights, wit, encouragement, and overall generosity.

ACKNOWLEDGMENTS

I would like to thank the faculty, staff, and administration of the University of the Sciences in Philadelphia for their kind support throughout this project. Thanks, in particular, to Paul Katz, Suzanne Murphy, Elia Eschenazi, Roberto Ramos, Peter Miller, Kevin Murphy, Sam Talcott, Justin Everett, and Jim Cummings for useful suggestions and encouragement.

I appreciate the kind help of the Wheeler and Feynman families, including James Wheeler, Letitia Wheeler Ufford, Alison Wheeler Lahnston, Carl Feynman, Michelle Feynman, and Joan Feynman—with thanks for the permission to publish selected personal correspondence. I am most grateful for the insightful comments and recollections of Freeman Dyson, Charles Misner, Virginia Trimble, Jayant Narlikar, Laurie Brown, Shirley Marneus, Cécile DeWitt-Morette, Kurt Gottfried, Kenneth Ford, and Linda Dalrymple Henderson. Thanks also to Betsy Devine, Frank Wilczek, Alan Chodos, Dean Rickles, and Chris DeWitt. I remain indebted to John Wheeler and Bryce DeWitt for interviews conducted in 2002 during my John Simon Guggenheim Memorial Fellowship.

Thanks to all those in the history of science, science writing, and literary communities who have encouraged my writing, including Michal Meyer, Robert Jantzen, Peter Pesic, J. David Jackson, Gregory Good, David Cassidy, Don Howard, Alex Wellerstein, Robert Romer, Joseph Martin, Cameron Reed, Robert Crease, Catherine Westfall, Marcus Chown, Graham Farmelo, Tasneem Zehra Husain, John Heilbron, Gerald Holton, Roger Stuewer, Gino Segre, Jo Alison Parker, Thony Christie, Kate Becker, Corey Powell, Ethan Siegel, Dave Goldberg, Peter Rose, Greg Lester, Mitchell and Wendy Kaltz, Mark Singer, Simone Zelitch, Doug Buchholz, Vance Lehmkuhl,

John Ashmead, Theodora Ashmead, David Zitarelli, Peter D. Smith, Roland Orzabal, Michael Gross, and Lisa Tenzin-Dolma. I'm grateful to Amanda Gefter for an inspiring lecture about Wheeler, delivered at the American Philosophical Society. Thanks to Victoria Carpenter for useful discussions and collaborations about the nature of time in Latin American literature.

I'm much obliged to M. Craig Getting and K. C. MacMillan of the Lantern Theater Company in Philadelphia for offering me the opportunity to serve as scientific consultant on a production of the play *QED* and inviting me to lead a postperformance discussion about Feynman. I'm also grateful to the Tyler Gallery of Temple University for inviting me to be a scientific consultant for an artistic event honoring John Wheeler. I am thankful to the American Philosophical Society for access to the John Wheeler Papers and to the Niels Bohr Library & Archives, American Institute of Physics, for access to its Oral Histories Collection, and to the Archives of the California Institute of Technology for access to the Papers of Richard Phillips Feynman.

This project would not have been possible without the extraordinary work of the editorial staff at Basic Books, including T. J. Kelleher, Hélène Barthélemy, Collin Tracy, Jen Kelland, and Quynh Do. I am grateful to my fantastic agent, Giles Anderson of the Anderson Literary Agency, for his help and suggestions for this project and his continued support in general.

I am most appreciative of the love and support of my family and friends, including my parents, Stanley and Bernice Halpern, and my in-laws Joseph and Arlene Finston, Shara Evans, Lane Hurewitz, and Jill Bernstein. Thanks to Michael Erlich and Fred Schuepfer for their friendly encouragement. Above all, many thanks to my wife, Felicia, for her valuable, insightful suggestions and to my sons, Eli and Aden, for their inspiring creativity.

NOTES

INTRODUCTION

7 **"Through some wonderful freak of fate":** John A. Wheeler, quoted in Christopher Sykes, ed., *No Ordinary Genius: The Illustrated Richard Feynman* (New York: W. W. Norton & Company, 1994), 44.

8 **"I was very lucky when I got to Princeton":** Richard P. Feynman, quoted in Dick Stanley, "A Pioneer of Thought," *Austin American-Statesman*, February 8, 1987.

8 **"Princeton has a certain elegance":** Interview of Richard Feynman by Charles Weiner on March 5, 1966, Niels Bohr Library & Archives, American Institute of Physics (AIP), College Park, MD, www.aip.org/history-programs/niels-bohr-library/oral-histories/5020-1.

9 **"I'll have both, thank you":** Richard P. Feynman, *Classic Feynman: All the Adventures of a Curious Character*, ed. Ralph Leighton (New York: W. W. Norton, 2006), 60.

9 **"as though Groucho Marx was suddenly standing":** C. P. Snow, *The Physicists* (Boston: Little Brown and Company, 1981), 143.

10 **A first-place finish in a New York University:** "Prizes Awarded in Mathematics," *New York Times*, May 19, 1935.

10 **"What makes it go?":** Feynman, *Classic Feynman*, 325.

10 **"Is the photon in the atom ahead of time":** Melville Feynman, reported by Richard Feynman in Sykes, *No Ordinary Genius*, 39.

11 **"He can always look at something the way a child does":** Ralph Leighton, interview in Warren E. Leary, "Puzzles Propel Physicist with Penchant for Probing," *Sunday Telegraph*, April 13, 1986, G6.

12 **"If I keep going out into space":** John Wheeler, quoted in John Boslough, "Inside the Mind of John Wheeler," *Reader's Digest*, September 1986, 107.

12 **Finding countless ways to tinker:** James Gleick, *Genius: The Life and Science of Richard Feynman* (New York: Vintage, 1993), 27.

12 **He experimented with gunpowder:** John Archibald Wheeler with Kenneth W. Ford, *Geons, Black Holes, and Quantum Foam: A Life in Physics* (New York: W. W. Norton & Company, 2000), 81–82.

CHAPTER ONE

18 **"Bohr has this probing approach to everything":** Interview of John Wheeler by Thomas S. Kuhn and John L. Heilbron on March 24, 1962, Niels Bohr Library & Archives, AIP, www.aip.org/history-programs/niels-bohr-library/oral-histories/4957.

21 **"Look, we have to get serious here":** Interview of Richard Feynman by Charles Weiner on March 5, 1966, Niels Bohr Library & Archives, AIP, www.aip.org/history-programs/niels-bohr-library/oral-histories/5020-1.

22 **"nicely and smoothly":** Richard P. Feynman to Lucille Feynman, October 11, 1939, in Richard P. Feynman, *Perfectly Reasonable Deviations (from the Beaten Track)*, ed. Michelle Feynman (New York: Basic Books, 2006), 2.

22 **"I don't know how to think without pictures":** John A. Wheeler, interview by the author in his Princeton office, November 5, 2002.

25 **"Wheeler was very much influenced by Niels Bohr":** Charles W. Misner, phone interview by the author, December 6, 2015.

31 **The cyclotron was in the middle of the room:** Interview of Richard Feynman by Charles Weiner on March 5, 1966, Niels Bohr Library & Archives, AIP, www.aip.org/history-programs/niels-bohr-library/oral-histories/5020-1.

33 **Feynman distracted himself during the boring lectures:** Richard P. Feynman, *Classic Feynman: All the Adventures of a Curious Character*, ed. Ralph Leighton (New York: W. W. Norton & Company, 2006), 60, 44.

37 **As a little girl, Joan had assisted Richard:** Christopher Riley, "Joan Feynman: From Auroras to Anthropology," in *A Passion for Science: Tales of Discovery and Invention*, ed. Suw Charman-Anderson (London: FindingAda, 2015).

38 **"I had no contact with Wheeler at all":** Joan Feynman, voicemail correspondence with the author, December 23, 2015.

38 **"I have a problem for you":** James Wheeler, phone interview by the author, October 31, 2015.

39 **"I have an image of Feynman":** Letitia Wheeler Ufford, phone interview by the author, October 31, 2015.

39 **"Niels Bohr sat in my mother's favorite red velvet club chair":** Alison Wheeler Lahnston, phone interview by the author, October 31, 2015.

CHAPTER TWO

50 **"space by itself and time by itself are doomed":** Hermann Minkowski, address delivered at the 80th Assembly of German Natural Scientists and Physicians, September 21, 1908.

52 **"To us believing physicists, the distinction":** Albert Einstein to Vero and Bice Besso, March 21, 1955, quoted in Albrecht Fölsing, *Albert Einstein*, trans. Ewald Osers (New York: Penguin, 1997), 741.

63 **"Hello, I'm coming to your seminar":** Reported by Richard P. Feynman in Ralph Leighton, ed., *Classic Feynman: All the Adventures of a Curious Character* (New York: W. W. Norton & Company, 2006), 67.

67 **"I'm awfully happy . . . that you're going to publish":** Arline Greenbaum to Richard P. Feynman, in Richard P. Feynman, *Perfectly Reasonable Deviations (from the Beaten Track)*, ed. Michelle Feynman (New York: Basic Books, 2006), 7.

72 **"The Lagrangian in Quantum Mechanics":** Paul A. M. Dirac, "The Lagrangian in Quantum Mechanics," *Physikalische Zeitschrift der Sowjetunion* 3, no. 1 (1933): 64–72. Reprinted in Laurie Brown, ed., *Feynman's Thesis: A New Approach to Quantum Theory* (Singapore: World Scientific, 2005), 113–121.

75 **"Wheeler told me that he—ever in search":** Kenneth W. Ford, correspondence with the author, December 28, 2015.

76 **"I can't believe that God plays dice":** John A. Wheeler, interview by the author in his Princeton office, November 5, 2002.

78 **"I have never been too busy to dream":** John Archibald Wheeler with Kenneth W. Ford, *Geons, Black Holes, and Quantum Foam: A Life in Physics* (New York: W. W. Norton & Company, 2000), 182.

CHAPTER THREE

88 **The Garden of Forking Paths is an incomplete:** Jorge Luis Borges, "The Garden of Forking Paths," in *Labyrinths: Selected Stories and Other Writings*, trans. James E. Irby (New York: New Directions, 1962), 28.

92 **"Both Wigner and Ladenburg feel with me":** John A. Wheeler to Richard P. Feynman, March 26, 1942, Papers of Richard Phillips Feynman, Archives, California Institute of Technology.

97 **"I realized very quickly that he was something phenomenal":** Hans Bethe, quoted in Jeremy Bernstein, *Hans Bethe: Prophet of Energy* (New York: Basic Books, 1979), 61.

97 **"No, no, you're crazy!":** Stephane Groueff, *Manhattan Project: The Untold Story of the Making of the Atomic Bomb* (Boston: Little Brown and Co., 1967), 202.

97 **One day Feynman, as a joke:** Lee Edson, "Scientific Man for All Seasons," *New York Times*, March 10, 1966.

98 **"He is a second Dirac":** Robert Oppenheimer to Raymond Birge, November 1943, quoted in Silvan S. Schweber, *QED and the Men Who Made It: Dyson, Feynman, Schwinger, and Tomonaga* (Princeton, NJ: Princeton University Press, 1994), 398–399.

100 **"Feynman seemed to be composed in equal parts":** Edward Teller, *Memoirs: A Twentieth-Century Journey in Science and Politics* (Cambridge, MA: Perseus, 2001), 168.

100 **"played them for hours each night":** Ibid.

102 **After discussing the results with the plant manager:** Interviews of John Wheeler by Kenneth W. Ford on October 5, 1994–April 12, 1995, Niels Bohr Library & Archives, AIP, www.aip.org/history-programs/niels-bohr-library/oral-histories/5908-13-22.

103 **"Even to the big shot guys":** Richard P. Feynman in Ralph Leighton, ed., *Classic Feynman: All the Adventures of a Curious Character* (New York: W. W. Norton & Company, 2006), 149.

105 **"We must, therefore, be prepared to find":** Niels Bohr, *Atomic Theory and the Description of Nature* (Cambridge: Cambridge University Press, 1934).

106 **"The Sun would not radiate":** H. Tetrode, "Über den Wirkungszusammenhang der Welt. Eine Erweiterung der Klassischen Dynamik," *Zeitschrift für Physik* 10 (1922): 317.

107 **"Pre-acceleration and the force of radiative reaction":** John A. Wheeler and Richard P. Feynman, "Interaction with the Absorber as the Mechanism of Radiation," *Reviews of Modern Physics* 17, nos. 2–3 (April–July 1945): 180–181.

110 **"I am convinced that the United States":** John Archibald Wheeler with Kenneth W. Ford, *Geons, Black Holes, and Quantum Foam: A Life in Physics* (New York: W. W. Norton & Company, 2000), 19.

112 **The clock was faulty:** Richard P. Feynman in Christopher Sykes, ed., *No Ordinary Genius: The Illustrated Richard Feynman* (New York: W. W. Norton & Co, 1994), 55.

CHAPTER FOUR

116 **"the 'idea-woman'":** Richard Feynman to Arline Feynman, October 17, 1946, reprinted in Richard P. Feynman, *Perfectly Reasonable Deviations (from the Beaten Track)*, ed. Michelle Feynman (New York: Basic Books, 2006), 69.

117 **"I love my wife. My wife is dead":** Ibid.

118 **"There grew up in time, to be sure":** Interviews of John Wheeler by Kenneth W. Ford on October 5, 1994–April 12, 1995, Niels Bohr Library & Archives, AIP, www.aip.org/history-programs/niels-bohr-library/oral-histories/5908-13-22

120 **"More and more the possibility suggests itself to us":** John A. Wheeler, reported in William Laurence, "'Super' Uranium Fission Held Possible; 50% Stronger Than Present Atomic Bomb," *New York Times*, December 3, 1947.

121 **"What will the fundamental particles turn out to be":** Richard P. Feynman, reported in G. T. Reynolds and Donald R. Hamilton, eds., *The Future of Nuclear Science* [Summary prepared by D. R. Hamilton,

Princeton University Bicentennial Conference on the Future of Nuclear Science] (Princeton, NJ: Princeton University Press, 1946).

121 **One topic of discussion at the conference:** Malcolm Browne, "Physicists Predict Progress in Solving Problem of Gravity," *New York Times*, November 5, 1996.

123 **"Listen, Buddy, the room situation is tough":** Interview of Richard Feynman by Charles Weiner on June 27, 1966, Niels Bohr Library & Archives, AIP, www.aip.org/history-programs/niels-bohr-library/oral-histories/5020-3.

127 **Brillouin, in turn, suggested asking:** Silvan S. Schweber, *QED and the Men Who Made It: Dyson, Feynman, Schwinger, and Tomonaga* (Princeton, NJ: Princeton University Press, 1994), 160.

131 **"Your work on the fine structure led directly":** Freeman Dyson to Willis Lamb, reported in Schweber, *QED and the Men Who Made It*, 218–219.

CHAPTER FIVE

134 **He had either missed the bus:** Interview of Richard Feynman by Charles Weiner on June 27, 1966, Niels Bohr Library & Archives, AIP, www.aip.org/history-programs/niels-bohr-library/oral-histories/5020-3.

138 **"Viki . . . felt that Bethe received too much recognition":** Kurt Gottfried, correspondence with the author, May 20, 2016.

141 **"All the mathematical proofs were later discoveries":** Richard P. Feynman to Ted Welton, November 19, 1949, quoted in David Kaiser, *Drawing Theories Apart: The Dispersion of Feynman Diagrams in Postwar Physics* (Chicago: University of Chicago Press, 2005), 178.

142 **"Feynman believed fervently that the diagrams":** Kaiser, *Drawing Theories Apart*, 177.

142 **Nevertheless, they were baffled:** Freeman J. Dyson, *Disturbing the Universe* (New York: Basic Books, 1981), 13.

143 **Dyson was also taken aback:** Ibid., 47.

143 **"Hearing about the sum over histories directly":** Freeman Dyson, correspondence with the author, December 19, 2015.

144 **"a magician of the highest caliber":** Mark Kac, *Enigmas of Chance: An Autobiography* (New York: Harper & Row, 1985), xxv.

148 **"Feynman had a completely new way of looking at things":** Hans Bethe, oral history interview by Judith R. Goodstein, February 17, 1982, Archives, California Institute of Technology.

149 **With hopes of visiting her:** Ralph Leighton, ed., *Classic Feynman: All the Adventures of a Curious Character* (New York: W. W. Norton & Company, 2006), 200.

154 **"After moving to Princeton in 1948":** Dyson correspondence, December 19, 2015.

154 **"It seemed that path integrals were an extremely powerful tool":** Barry Simon, *Functional Integration and Quantum Physics* (New York: Academic Press, 1979), preface, quoted in Cécile DeWitt-Morette, *The Pursuit of Quantum Gravity: Memoirs of Bryce DeWitt from 1946 to 2004* (New York: Springer Verlag, 2011), 13.

155 **"That afternoon, Feynman produced more brilliant ideas":** Freeman Dyson, "Of Historical Note: Richard Feynman," Institute for Advanced Study, http://www.ias.edu/ideas/2011/dyson-of-historical-note.

156 **"The sum over histories finally gave us":** Dyson correspondence, December 19, 2015.

156 **"Well, I probably saw him a total of 20 minutes altogether":** Bryce DeWitt, phone interview by the author, December 4, 2002.

157 **After a long, fun day canoeing together:** Cécile DeWitt-Morette, "Snapshots," Institute for Advanced Study, http://www.ias.edu/ideas/2011-dewitt-morette-ias.

159 **Before walking up to him:** John Archibald Wheeler with Kenneth W. Ford, *Geons, Black Holes, and Quantum Foam: A Life in Physics* (New York: W. W. Norton & Company, 2000), 188.

161 **most of it is now common knowledge:** Kenneth W. Ford, *Building the Bomb: A Personal History* (Singapore: World Scientific, 2015), 1.

162 **"I know you plan to spend next year":** John A. Wheeler to Richard P. Feynman, March 29, 1951, reprinted in Richard P. Feynman, *Perfectly Reasonable Deviations (from the Beaten Track)*, ed. Michelle Feynman (New York: Basic Books, 2006), 83.

165 **"I wanted to know your opinion of our old theory":** Richard P. Feynman to John A. Wheeler, May 4, 1951, Papers of Richard Phillips Feynman, Archives, California Institute of Technology.

165 **"The idea [of direct interactions between electrons] seemed so obvious":** Richard P. Feynman, "Nobel Lecture," Nobelprize.org, December 11, 1965, http://www.nobelprize.org/nobel_prizes/physics/laureates/1965/feynman-lecture.html (accessed June 25, 2016).

CHAPTER SIX

168 **He called his approach "radical conservative-ism":** Charles W. Misner, Kip S. Thorne, and Wojciech H. Zurek, "John Wheeler, Relativity, and Quantum Information," *Physics Today* (April 2009): 40.

169 **"We live on an island surrounded by a sea of ignorance":** John A. Wheeler, quoted in John Horgan, "Gravity Quantized?," *Scientific American* 267, no. 3 (September 1992): 18–19.

169 **"I think I can safely say that nobody":** Richard P. Feynman, "The Character of Physical Law," BBC Television, 1965.

170 **"My father spoke about the warp":** Alison Wheeler Lahnston, phone interview by the author, October 31, 2015.

172 "playground for mathematicians": John Archibald Wheeler with Kenneth W. Ford, *Geons, Black Holes, and Quantum Foam: A Life in Physics* (New York: W. W. Norton & Company, 2000), 232.

174 "The shell game that we play": Richard P. Feynman, *QED: The Strange Theory of Light and Matter* (Princeton, NJ: Princeton University Press, 1985), 128.

175 "He begins working calculus problems in his head": "Hubby Got Custody of African Drums," *Star-News*, July 18, 1956.

176 "This house will never become a place of pilgrimage": Oscar Wallace Greenberg, "Visits with Einstein and Discovering Color in Quarks: Memories of the Institute for Advanced Study," Institute for Advanced Study, 2015, https://www.ias.edu/ideas/2015/greenberg-color (accessed January 31, 2017).

177 "Space acts on matter": Charles W. Misner, Kip S. Thorne, and John A. Wheeler, *Gravitation* (San Francisco: W. H. Freeman, 1973), 5.

180 "This sum over histories idea for doing quantum gravity": Interview of Charles Misner by Christopher Smeenk on May 22, 2001, Niels Bohr Library & Archives, AIP, www.aip.org/history-programs/niels-bohr-library/oral-histories/33697.

180 "reality proceeds by an awareness of all the possibilities": Charles W. Misner, phone interview by the author, December 6, 2015.

182 The brainy quartet often hung out in Everett's room: Peter Byrne, *The Many Worlds of Hugh Everett III: Multiple Universes, Mutual Assured Destruction, and the Meltdown of a Nuclear Family* (New York: Oxford University Press, 2013), 57.

182 "highest of its kind and next only": "Einstein Award to Professor," *New York Times*, March 14, 1954.

185 "Hugh thought that Petersen's interpretation was intolerable": Misner, phone interview, December 6, 2015.

187 "My first and general reaction was": Ibid.

188 Feynman replied on October 4 with a brief: Richard P. Feynman to John A. Wheeler, October 4, 1955, Wheeler Archive, American Philosophical Society.

189 "There's a certain irrationality to any work": Richard P. Feynman, "Quantum Theory of Gravitation," *Acta Physica Polonica* 24 (1963): 267.

190 "Probability in Wave Mechanics": Byrne, *Many Worlds of Hugh Everett*, 138.

190 "As soon as the observation is performed": Hugh Everett III, "The Theory of the Universal Wave Function," draft of PhD thesis, Princeton University, reprinted in Bryce DeWitt and Neill Graham, eds., *The Many-Worlds Interpretation of Quantum Mechanics* (Princeton, NJ: Princeton University Press, 1973), 98–99.

194 "[It] was a conference by invitation only": Bryce DeWitt, phone interview by the author, December 4, 2002.

194 **When he arrived at the Raleigh-Durham airport:** Ralph Leighton, ed., *Classic Feynman: All the Adventures of a Curious Character* (New York: W. W. Norton & Company, 2006), 273.

194 **"Feynman showed up and he said, 'Hi Geon'":** DeWitt, phone interview, December 4, 2002.

194 **"Nobody believed in it":** Ibid.

196 **"I was so shocked that I sat down":** Bryce DeWitt, reported in Cécile DeWitt-Morette, *The Pursuit of Quantum Gravity: Memoirs of Bryce DeWitt from 1946 to 2004* (New York: Springer Verlag, 2011), 94.

197 **"The concept of a 'universal wave function'":** Richard P. Feynman, reported in Cécile M. DeWitt and Dean Rickles, eds., *The Role of Gravitation in Physics: Report from the 1957 Chapel Hill Conference* (Berlin: Edition Open Access, 2011), 270.

197 **"Feynman had no use for philosophy":** Freeman Dyson, correspondence with the author, December 19, 2015.

197 **"I do not remember when I first heard":** Ibid.

CHAPTER SEVEN

203 **"scheme for pushing a great problem under the rug":** Richard P. Feynman, quoted in "Caltech Nobel Winner Modest on Findings," *Los Angeles Times*, October 22, 1965, 2.

205 **"As I thought about it, as I beheld it in my mind's eye":** Richard P. Feynman, interview by Jagdish Mehra, Pasadena, California, January 1988, reported in Jagdish Mehra, *The Beat of a Different Drum: The Life and Science of Richard Feynman* (New York: Oxford University Press, 1994), 453.

207 **"A lecture by Dr. Feynman is a rare treat indeed":** Irving S. Bengelsdorf, "Caltech's Feynman Brings Artist's Touch to Physics," *Los Angeles Times*, March 14, 1967, A.

207 **"Ask me anything":** James Cummings, conversation with the author, January 25, 2016.

207 **Some of the freshmen called it the "Oracle of Feynman":** "Groveling frosh humbly seeks physics inspiration from Oracle of Feynman at Dabney," photo caption, *California Tech*, October 28, 1965, 1.

209 **"Feynman was quite explicit on this":** Jayant Narlikar, correspondence with the author, January 9, 2016.

210 **"I recall the talk given by Dennis Sciama":** Ibid.

212 **"I got a telephone call from [Wheeler]":** Bryce DeWitt, phone interview by the author, December 4, 2002.

212 **"The universe is not a system":** John A. Wheeler, "Three-Dimensional Geometry as a Carrier of Information About Time," in *The Nature of Time*, ed. Thomas Gold with the assistance of D. L. Schumacher (Ithaca, NY: Cornell University Press, 1967), 106–107.

213 **"He started by being against field theory":** Narlikar, correspondence, January 9, 2016.

213 **"The new theory draws on the mathematical reasoning":** Walter Sullivan, "Scientist Revises Einstein's Theory," *New York Times*, June 21, 1964.

214 **"Two observers who were able to talk":** Charles W. Misner, "Infinite Red-Shifts in General Relativity," in *The Nature of Time*, ed. Thomas Gold with the assistance of D. L. Schumacher (Ithaca, NY: Cornell University Press, 1967), 75.

215 **Several magazines covering the event:** Tom Siegfried, "50 Years Later, It's Hard to Say Who Named Black Holes," *Science News*, December 23, 2013, http://www.sciencenews.org/blog/context/50-years-later-it's-hard-say-who-named-black-holes (accessed June 24, 2016).

218 **"for their fundamental work in quantum electrodynamics":** "The Nobel Prize in Physics 1965," Nobelprize.org, http://www.nobelprize.org/nobel_prizes/physics/laureates/1965 (accessed June 25, 2016).

218 **"Feynman spotted me crossing Caltech campus one day":** Virginia Trimble, correspondence with the author, February 10, 2017.

219 **"Feynman came to my office around eight a.m.":** Virginia Trimble, correspondence with the author, February 9, 2017.

220 **But when he got up on stage dressed formally:** Mehra, *Beat of a Different Drum*, 576–577.

221 **"Although I did do many of the things described":** Richard P. Feynman, letter to the editor, *New York Times*, November 5, 1967.

225 **"the three gates of time":** Linda Anthony, "The Big Bang . . . the Big Crunch," *Austin American-Statesman*, May 20, 1979, C15, archived in Wheeler Papers, American Philosophical Society.

225 **At a 1966 American Physical Society meeting, he speculated:** Walter Sullivan, "Physicists Muse on Question of Time Running Backward," *New York Times*, January 30, 1966.

CHAPTER EIGHT

232 **One day, Wheeler was joking with Bekenstein:** John Archibald Wheeler with Kenneth W. Ford, *Geons, Black Holes, and Quantum Foam: A Life in Physics* (New York: W. W. Norton & Company, 2000), 314.

234 **"This guy sounds crazy":** Charles W. Misner, Kip S. Thorne, and Wojciech H. Zurek, "John Wheeler, Relativity, and Quantum Information," *Physics Today* (April 2009): 44–45.

236 **"Rasputin, Science, and the Transmogrification of Destiny":** Charles W. Misner, "John Wheeler and the Recertification of General Relativity as True Physics," in *General Relativity and John Archibald Wheeler*, ed. I. Ciufolini and R. A. Matzner. Astrophysics and Space Science Library (New York: Springer Verlag, 2010).

236 **"Let the universe have a destiny":** John A. Wyler (pseudonym), "Rasputin, Science, and the Transmogrification of Destiny," *General Relativity and Gravitation* 5, no. 2 (1974): 176–177.

237 **Fredkin and Minsky met Feynman for the first time on a lark:** Julian Brown, *Minds, Machines, and the Multiverse: The Quest for the Quantum Computer* (New York: Simon and Schuster, 2000), 60.

239 **They'd laugh and laugh at their own clever silliness:** "An Interview with Michelle Feynman," Basic Feynman, 2005, http://www.basicfeynman.com/qa.html.

239 **Much to Feynman's consternation:** Freeman Dyson, personal recollection to the author, Institute for Advanced Study, December 9, 2016.

239 **"I loved *Gravitation* as a child and was chuffed":** Carl Feynman, correspondence with the author, July 24, 2016.

240 **"What ever happened to Tannu Tuva?":** Richard P. Feynman, reported in Ralph Leighton, *Tuva or Bust! Richard Feynman's Last Journey* (New York: W. W. Norton & Company, 1991), 248.

240 **Wheeler asked Feynman to contribute to a panel:** John A. Wheeler to Richard P. Feynman, June 1978, Wheeler Papers, American Philosophical Society.

240 **Feynman politely declined and jokingly reported:** Richard P. Feynman to John A. Wheeler, June 14, 1978, Wheeler Papers, American Philosophical Society.

240 **"Welcome home," he wrote:** John A. Wheeler to Richard P. Feynman, July 28, 1978, Wheeler Papers, American Philosophical Society.

243 **this would be Feynman and Dyson's last meeting:** Dyson, personal recollection, December 9, 2016.

244 **"I'm a big fool, but I enjoy life":** Richard P. Feynman, quoted in Dick Stanley, "A Pioneer of Thought," *Austin American-Statesman*, February 8, 1987.

244 **"One of the biggest regrets of my life":** Ibid.

246 **There was a moment of shocked silence:** Shirley Marneus, phone interview by the author, February 21, 2017.

248 **"A year ago I would have told you":** Richard P. Feynman, quoted in David E. Sanger, "A Computer Full of Surprises," *New York Times*, May 8, 1987.

249 **"We should be looking at broader questions":** John A. Wheeler, interview by the author in his Princeton office, November 5, 2002.

249 **"I had noticed when I was younger":** Richard P. Feynman, reported in P. C. W. Davies and J. Brown, eds., *Superstrings: A Theory of Everything?* (New York: Cambridge University Press, 1988), 193.

249 **"Quantum Theory, the Church-Turing Principle and the Universal Quantum Computer":** David Deutsch, "Quantum Theory, the Church-Turing Principle and the Universal Quantum Computer," *Proceedings of the Royal Society of London* A400 (1985): 97–117.

250 **"Philosophy is too important to be left":** John A. Wheeler, reported in Dwight E. Neuenschwander, ed., "The Scientific Legacy of John Wheeler," *APS Forum on the History of Physics Newsletter*, fall

2009, https://www.aps.org/units/fhp/newsletters/fall2009/wheeler.cfm (accessed July 3, 2016).

251 **"The universe does not exist 'out there'":** John A. Wheeler, "The Participatory Universe," *Science* 81 (June 1981): 66–67.

252 **"All I'm interested in":** Richard P. Feynman, reported in Davies and Brown, *Superstrings*, 203.

253 **"Physics was exciting then":** John A. Wheeler to Richard P. Feynman, June 27, 1986, Wheeler Papers, American Philosophical Society.

253 **"For a successful technology, reality must take precedence":** Richard P. Feynman, "Personal Observations on the Reliability of the Shuttle," Appendix F, Rogers Commission Report, NASA (1986), http://science.ksc.nasa.gov/shuttle/missions/51-l/docs/rogers-commission/Appendix-F.txt (accessed January 31, 2017).

253 **"Didn't I inherit from you the faculty":** John A. Wheeler to Richard P. Feynman, August 7, 1986, Wheeler Papers, American Philosophical Society.

253 **He complained vehemently:** John Archibald Wheeler, "A Decade of Permissiveness," *New York Review of Books*, May 17, 1979, 41–44.

257 **"Tycho Brahe had his supernova":** David L. Goodstein and Gerry Neugebauer, special preface, in Richard P. Feynman with Robert B. Leighton and Matthew Sands, *Six Not-So-Easy Pieces: Einstein's Relativity, Symmetry, and Space-Time* (New York: Basic Books, 2011), xviii.

258 **the vice president of the Soviet Academy of Sciences had just mailed him an invitation:** Jagdish Mehra, *The Beat of a Different Drum: The Life and Science of Richard Feynman* (New York: Oxford University Press, 1994), 606.

258 **"I'd hate to die twice":** Richard P. Feynman, *Perfectly Reasonable Deviations (from the Beaten Track)*, ed. Michelle Feynman (New York: Basic Books, 2006), 373.

260 **"The poetic Wheeler is a prophet":** Freeman Dyson, quoted in Dennis Overbye, "John A. Wheeler, Physicist Who Coined the Term 'Black Hole,' Is Dead at 96," *New York Times*, April 14, 2008.

CONCLUSION

263 **"The game I play is a very interesting one":** Richard P. Feynman, quoted in Christopher Sykes, ed., *No Ordinary Genius: The Illustrated Richard Feynman* (New York: W. W. Norton & Company, 1994), 98.

EPILOGUE

270 **Wheeler sat in the front of the large lecture:** For a fascinating perspective on Wheeler's ninetieth birthday celebration, see Amanda

Gefter, *Trespassing on Einstein's Lawn: A Father, a Daughter, the Meaning of Nothing, and the Beginning of Everything* (New York: Bantam, 2014).

270 **"Well here's your chance to go on":** John A. Wheeler, interview by the author in his Princeton office, November 5, 2002.

FURTHER READING

Bartusiak, Marcia. *Black Hole: How an Idea Abandoned by Newtonians, Hated by Einstein, and Gambled On by Hawking Became Loved.* New Haven, CT: Yale University Press, 2015.

Bernstein, Jeremy. "What Happens at the End of Things?," *Alcade* 74 (November/December 1985) 4–12.

Boslough, John. "Inside the Mind of John Wheeler." *Reader's Digest* (September 1986): 106–110.

Brown, Julian. *Minds, Machines, and the Multiverse: The Quest for the Quantum Computer.* New York: Simon and Schuster, 2000.

Brown, Laurie M., and John S. Rigden, eds. *"Most of the Good Stuff": Memories of Richard Feynman.* Washington, DC: American Institute of Physics, 1993.

Byrne, Peter. *The Many Worlds of Hugh Everett III: Multiple Universes, Mutual Assured Destruction, and the Meltdown of a Nuclear Family.* New York: Oxford University Press, 2013.

Carpenter, Victoria, and Paul Halpern. "Quantum Mechanics and Literature: An Analysis of El Túnel by Ernesto Sábato." *Ometeca* 17 (2012), 167–187.

DeWitt-Morette, Cécile. *The Pursuit of Quantum Gravity: Memoirs of Bryce DeWitt from 1946 to 2004.* New York: Springer, 2011.

Dirac, Paul. *The Principles of Quantum Mechanics.* Oxford: Oxford University Press, 1930.

Dresden, Max. *H. A. Kramers: Between Tradition and Revolution.* New York: Springer Verlag, 1987.

Dyson, Freeman J. *Disturbing the Universe.* New York: Basic Books, 1981.

Everett, Justin, and Paul Halpern. "Spacetime as a Multicursal Labyrinth in Literature with Application to Philip K. Dick's *The Man in the High Castle.*" *KronoScope* 13, no. 1 (2013).

Farmelo, Graham. *The Strangest Man: The Hidden Life of Paul Dirac, Mystic of the Atom.* New York: Basic Books, 2009.

Feynman, Richard P. *Classic Feynman: All the Adventures of a Curious Character,* ed. Ralph Leighton. New York: W. W. Norton & Company, 2006.

———. *Feynman's Thesis: A New Approach to Quantum Theory*, ed. Laurie M. Brown. Singapore: World Scientific, 2005.

———. *Perfectly Reasonable Deviations (from the Beaten Track)*. ed. Michelle Feynman. New York: Basic Books, 2006.

———. *QED: The Strange Theory of Light and Matter*. Princeton, NJ: Princeton University Press, 1985.

———. *The Quotable Feynman*, ed. Michelle Feynman. Princeton, NJ: Princeton University Press, 2015.

———. *"What Do You Care What Other People Think?": Further Adventures of a Curious Character*, ed. Ralph Leighton. New York: W. W. Norton & Company, 2001.

Feynman, Richard P., with Ralph Leighton. *Surely You're Joking, Mr. Feynman! Adventures of a Curious Character*, ed. Edward Hutchings. New York: W. W. Norton & Company, 1997.

Ford, Kenneth W. *Building the Bomb: A Personal History*. Singapore: World Scientific, 2015.

Gefter, Amanda. "Haunted by His Brother, He Revolutionized Physics." *Nautilus*, January 16, 2014, http://nautilus/issue/9/time/haunted-by-his-brother-he-revolutionized-physics.

———. *Trespassing on Einstein's Lawn: A Father, a Daughter, the Meaning of Nothing, and the Beginning of Everything*. New York: Bantam, 2014.

Gleick, James. *Genius: The Life and Science of Richard Feynman*. New York: Vintage, 1993.

Gribbin, John, with Mary Gribbin. *Richard Feynman: A Life in Science*. London: Penguin Books, 1997.

Halliwell, J. J., J. Perez-Mercader, and W. H. Zurek, eds. *The Physical Origins of Time-Asymmetry*. Cambridge: Cambridge University Press, 1996.

Halpern, Paul. *Einstein's Dice and Schrödinger's Cat: How Two Great Minds Battled Quantum Randomness to Create a Unified Theory of Physics*. New York: Basic Books, 2015.

———. *The Great Beyond: Higher Dimensions, Parallel Universes, and the Extraordinary Search for a Theory of Everything*. Hoboken, NJ: Wiley, 2004.

———. "Time as an Expanding Labyrinth of Information." *KronoScope* 10, nos. 1–2 (2010): 64–76.

———. *Time Journeys: A Search for Cosmic Destiny and Meaning*. New York: McGraw-Hill, 1990.

Husain, Tasneem Zehra. *Only the Longest Threads*. Philadelphia: Paul Dry Books, 2014.

Kaiser, David, *Drawing Theories Apart: The Dispersion of Feynman Diagrams in Postwar Physics*. Chicago: University of Chicago Press, 2005.

Krauss, Lawrence M. *Quantum Man: Richard Feynman's Life in Science*. New York: W. W. Norton & Company, 2012.

Leighton, Ralph. *Tuva or Bust! Richard Feynman's Last Journey*. New York: W. W. Norton & Company, 1991.

Mach, Ernst. *The Science of Mechanics: A Critical and Historical Exposition of Its Principles*, trans. Thomas McCormack. Chicago: Open Court, 1897.

Mehra, Jagdish. *The Beat of a Different Drum: The Life and Science of Richard Feynman*. New York: Oxford University Press, 1994.

Misner, Charles W., Kip S. Thorne, and John A. Wheeler. *Gravitation*. San Francisco: W. H. Freeman, 1973.

Mlodinow, Leonard. *Feynman's Rainbow: A Search for Beauty in Physics and in Life*. New York: Vintage, 2011.

Schweber, Silvan S. *QED and the Men Who Made It: Dyson, Feynman, Schwinger, and Tomonaga*. Princeton, NJ: Princeton University Press, 1994.

Sykes, Christopher, ed. *No Ordinary Genius: The Illustrated Richard Feynman*. New York: W. W. Norton & Company, 1994.

Weisskopf, Victor. *The Joy of Insight: Passions of a Physicist*. New York: Basic Books, 1991.

Wheeler, John Archibald. "Time Today." In *The Physical Origins of Time-Asymmetry*, edited by J. J. Halliwell, J. Perez-Mercader, and W. H. Zurek, 1–29. Cambridge: Cambridge University Press, 1996.

Wheeler, John Archibald, with Kenneth W. Ford. *Geons, Black Holes, and Quantum Foam: A Life in Physics*. New York: W. W. Norton & Company, 2000.

Yourgrau, Palle. *A World Without Time: The Forgotten Legacy of Gödel and Einstein*. New York: Basic Books, 2004.

INDEX

University of the Sciences

PAUL HALPERN is professor of physics at the University of the Sciences in Philadelphia and author of fifteen popular science books, most recently *Einstein's Dice and Schrödinger's Cat*. He has received a Guggenheim Fellowship, a Fulbright Scholarship, and an Athenaeum Literary Award. Halpern has appeared on numerous radio and television shows, including *Future Quest*, *Radio Times*, several shows on the History Channel, and "The Simpsons 20th Anniversary Special." He has contributed opinion pieces for the *Philadelphia Inquirer* and was also a regular contributor to NOVA's *The Nature of Reality* physics blog. He lives near Philadelphia, Pennsylvania.